Scientific Controversies

Scientific Controversies

A Socio-Historical Perspective on the Advancement of Science

Dominique Raynaud

With a preface by Mario Bunge

Routledge
Taylor & Francis Group

LONDON AND NEW YORK

First published in paperback 2024

First published 2015 by Transaction Publishers

Published 2017 by Routledge
4 Park Square, Milton Park, Abingdon, Oxon OX14 4RN

and by Routledge
605 Third Avenue, New York, NY 10158

Routledge is an imprint of the Taylor & Francis Group, an informa business

Library of Congress Catalog Number: 2014024490

Library of Congress Cataloging-in-Publication Data

Raynaud, Dominique.
   [Sociologie des controverses scientifiques. English]
   Scientific controversies / Dominique Raynaud; translated from the
French by Lisa C. Chien.
       pages cm
   In English.
   Originally published in French: Sociologie des controverses
scientifiques (Paris: Presses universitaires de France, 2003).
   Includes bibliographical references and index.
   ISBN 978-1-4128-5571-6
   1. Science--Social aspects. 2. Science--Philosophy. I. Chien, Lisa C.,
translator. II. Title.
   Q175.5.R3813 2015
   501--dc23
                                                        2014024490

ISBN: 978-1-4128-5571-6 (hbk)
ISBN: 978-1-03-292230-0 (pbk)
ISBN: 978-1-315-12903-7 (ebk)

DOI: 10.4324/9781315129037

Contents

List of Illustrations

List of Tables

List of Tables

Foreword

I am pleased to present *Scientific Controversies: A Socio-Historical Perspective on the Advancement of Science*. This is an expanded version of the work in French, previously published by the Presses universitaires de France, and translated with the joint support of Grenoble University and PLC (Philosophie, Langages & Cognition). I have written a detailed historical chronicle for Chapter 3 (The Vitalism-Organicism Controversy between Paris and Montpellier). The book includes a new Chapter 5 (Al-Samarqandī's Native Theory of Controversies), a fascinating Arab theorist unknown to me until recently. Finally, I have elaborated further on my Epistemological Conclusions, to better reflect the role of incrementalism in the pursuit of truth.

I am thankful to Mario Bunge who agreed to write a Preface to this book, to Lisa C. Chien for her excellent translation, and to the Transaction Publishers team for commendably professional work.

D. Raynaud
December 30, 2014

Preface

Up until the 1950s, the study of scientific communities was a task for philosophers, sociologists and historians of science intent on finding truths about science, that much-celebrated yet still elusive animal. Suffice it to recall the studies by Henri Poincaré, Emile Meyerson, Federigo Enriques, the Vienna Circle, Karl Popper, Ernest Nagel, Richard Braithwaite, Aldo Mieli, George Sarton, and Robert K. Merton.

In his classic 1938 paper on "Science and the social order," in the new journal *Philosophy of Science*, Merton had argued that the peculiarities of basic science are disinterestedness, universality, epistemic communism, and organized skepticism—not the doubt of the isolated researcher but critical scrutiny by an entire community.

Unlike his later critics, Merton was not an improvised parachutist but the first professional sociologist of science. His teachers had been the leading sociologists of his day—Pitirim Sorokin, George Sarton, and Talcott Parsons—as well as the amazing chemist, biologist, and sociologist Lawrence J. Henderson, who had rescued and popularized the concept of a social system. Besides, thanks to his wife, Harriet Zuckermann, Merton got to personally know many Nobel laureates, who told him details about what had made them tick, how they had worked, and how their respective scientific communities had now stimulated, now inhibited them.

In sum, around 1950 Merton was recognized as the most learned member of the science-studies communities. He was also the most balanced of them. The only one who, though not an idealist, stressed the disinterestedness of basic research. While not a positivist, Merton admitted the continuity and cumulative nature of science; and, though not a Marxist, he admitted the social embeddedness of the scientific community, as well as the economic and political pressures it was subjected to.

Not everyone shared the most popular images at that time. Some thought they resembled the descriptions of the elephant proposed by the blind Indian sages of legend. However, any of their descriptions would have helped in finding the elephant that had escaped from the circus, for every one of them held a grain of truth. Yes, truth, that black beast of the postmodernists.

Suddenly, in 1962, the elephant and the blind Indian sages vanished. In *The Structure of Scientific Revolutions,* Thomas S. Kuhn, an obscure scholar, held that scientists do not seek truth because there is no such thing—nor is there a body of knowledge that grows and is being repaired and made ever deeper.

Kuhn's central thesis was that, once in a while, there are scientific revolutions that sweep away everything that preceded them. Moreover, such radical changes would not solve long-standing scientific problems, but they would respond to alterations in the *Zeitgeist* or cultural fashion of the day, which Wilhelm Dilthey had invoked. Hence, scientists would neither find nor confirm nor confute anything. As his friend and comrade in arms Paul Feyerabend had declared, anything goes.

In short, according to Kuhn and Feyerabend, the classical picture of scientific research as the search for truth would be in crisis. Science would be a matter of opinion and social change. Hence, any amateur with enough *chutzpah* qualified for a job in one of the many "science studies" centers or "science and society" programs that proliferated over the past few decades.

This counter-revolution was so massive and so sudden, that it took the academic community by surprise and by storm. For example, I witnessed how Karl Popper, in his memorable confrontation with Kuhn at Bedford College, in the summer of 1965, adopted a defensive tactic and attempted to belittle his differences with Kuhn. Popper held that he too had never been interested in what Kuhn called "normal science." He even attempted to win the good will of his interlocutor, whom he started to call Tom.

Karl did not ask Tom to exhibit examples of revolutions out of nothing and that replaced well-confirmed hypotheses that were also congruent with the bulk of antecedent knowledge. He could not resort to this tactic because of his radical skepticism and his contempt for the sociology of knowledge. Besides, Popper could not ask Tom for examples, because he had claimed that only counter-examples matter.

In his *Popper and After* (1982), the Australian philosopher David Stove showed that Popper's radical skepticism was close to the antiscientism of Kuhn, Feyerabend, and Lakatos. However, his critique was so irreverent, and the alternative Stove proposed—namely a return to common sense and inductivism—was so naive, that nobody took him seriously.

Initially I underrated the antiscience antics of Kuhn, Feyerabend and their French counterparts. I thought that they were a passing fad that had gained notoriety just because it surfed on the wave of mistrust of science, that the Berkeley rebels had just denounced as an accomplice of the warlike and antienvironmental establishment.

I also thought that the amateurs who paraded as science experts were making use of the trick that Diderot had suggested to Rousseau when he came to Paris in search of instant celebrity: attack the prevailing opinion, that the sciences and the arts advance civilization. Rousseau followed this cynical advice, and his *Discourse on the Sciences and Arts* (1750) gained him the fame he craved.

It took me two decades to realize the seriousness of the damage those cheeky amateurs had done to academia, as well as the reason for their popularity, namely that they flattered the many students and scholars who had chosen the wide door—that of cultural studies, gender studies, science marketing, and the like. Every revolution elicits a reaction from former elites. In particular, every enlightenment movement elicits an obscurantist reaction.

I finally reacted in my 1991 and 1992 papers in *Philosophy of the Social Sciences*, where I examined not only the main contemporary anti-Mertonians, but also some of their precursors, from Karl Marx to Gaston Bachelard and Ludwik Fleck—the first to claim that "scientific facts" are inventions of scientific communities. The young Kuhn had read this outrageous thesis, which Bruno Latour was to exploit with great success among people whose heads had already been pummeled by Louis Althusser and Michel Foucault.

In those papers I also traced the philosophical roots of the counter-revolution in question: the subjectivism of Berkeley and Kant, as well as Marx's sociologism (or externalism) of 1859, that held that "society thinks through the individual." Boris Hessen, the Russian "organic intellectual," as Gramsci might have called him, applied that thesis to Newton, and expounded it with great success at the Second International Congress of History of Science, held in London in 1931.

However, my papers were not as successful, by far, as the excellent *Impostures Intellectuelles* (1997) by the physicists Alan Sokal and Jean Bricmont, published in the wake of Sokal's 1996 hilarious hoax

"Transgressing the boundaries: Towards a transformative hermeneutics of quantum gravity," whose publication in *Social Text* put in evidence the ignorance of its editors.

The so-called *science wars* that had followed the enthusiasm and support for science generated by Sputnik are still being fought. Scientism, strong and prestigious one century ago, is now weak and discredited in the humanist camp. Merton's realistic image of basic science has all but been abandoned by the metascience students, both on the right and the left.

Scientism is also being reviled in the social studies, where the pronouncements of the conservative ideologues Friedrich Hayek, Leo Strauss, Carl Schmitt, Hannah Arendt, Samuel Huntington, and Francis Fukuyama are far better known than the sober studies of Bob Merton, his teacher George Sarton, or Joseph Needham, the biochemist turned historian of Chinese culture.

But suddenly, in Grenoble, far from glittering Paris, surfaces Dominique Raynaud, an architect with a PhD in sociology—a rare combination. Raynaud resumes the line of rigorous work of Robert Merton and his school. His meticulous case studies are not flashy, hence they are unlikely to attract the worship of the uneducated. Raynaud's is the silent but solid work of the potter who fashions beautiful amphoras out of formless clay—in this case that of historical and sociological data.

Most of Raynaud's numerous publications are careful studies of particular contributions to science and technology. His early work focused largely on the "geometrical sciences," as d'Alembert might have called them, from geometry to optics, perspective, and architecture. He laid great stress on the long-term duration of the scientific process: geometry and optics took two millennia to develop! For example, descriptive geometry evolved slowly through its progressive mathematization, from the medieval stone-cutting techniques to Gaspard Monge, the mathematician whom Napoleon appointed senator.

Raynaud's aim has been to find truths buried in archival material and highly specialized journals. Yes, Raynaud has sought truths, not the power that Foucault and his followers imagined lurking underneath science, which they regarded as "politics by other means."

Raynaud's work continues that of my late friends Robert Merton and Raymond Boudon, as well as that of his mentor, Joseph Ben David. In this book, Raynaud deals with scientific and philosophical controversies, such as the debate over spontaneous generation between Pasteur and Pouchet, and the controversy over the propagation of light, which had polarized medieval scholars.

Raynaud has also contributed a clear vision of the so-called *science wars* of the last few decades. To him, all of the scientific controversies have been about truth. Certainly, some of them have had ideological overtones and motivations. But none of them was terminated by arbitrary political decision. True, the Galileo/Inquisition and Genetics/Stalinism controversies involved extrascientific authorities, but only for a while: in the end, truth won out.

My own experience as a frequent scientific controversialist confirms Raynaud's thesis that such conflicts are over truth, not power. They all involved exclusively conceptual disagreements. In particular the view that quantum mechanics refers to mental events, not to autonomously existing physical objects, invites ferreting out the referents of the basic concepts of the theory rather than the *ex cathedra* opinions of their founders. The scientific standing of the rational choice theories requires checking whether their central concepts—those of subjective utility and subjective probability—are mathematically well defined, and whether experiment supports the hypothesis of maximizing behavior. And the worth of psychoanalysis can be assessed in both the psychological lab and clinic, as well as by checklisting its claims to scientific standing.

That truth, not power, was at stake in all of these cases, as well as in those discussed by Raynaud, is a point in favor of scientism—the thesis that whatever can be investigated is best studied using the scientific method. By the same token, it is also a point against the intuitionism inherent in the "humanistic" school, as well as against the sociologism typical of both the early Marxist and Durkhemian schools in the sociology of knowledge.

When doing research in sociology or historiography, Raynaud regards these disciplines as branches of science, not of *Geisteswissenschaften* (spiritual sciences) requiring a method of their own, such as the *Verstehen* or empathic understanding exalted by Wilhelm Dilthey in his 1883 anti-scientistic manifesto. To be sure, putting oneself in *A*'s position (no mean feat!) may help explain why *A* thought or did *B*, but it does not explain *B* itself. Likewise, placing *A* in his/her social context may help explain why *B* was either recognized or suppressed by the establishment but, again, it does not account for *B* itself.

Because of his adherence to the scientific method, Raynaud writes clearly and gives good reasons for his views. Not for him the issuing of shocking slogans such as *Anything goes!* or of groundless formulas such as *Science is politics by other means.* Raynaud seeks to construct, not to deconstruct, as well as to clarify, not to obscure. He is a careful scholar, not an ideologist in search of instant celebrity. Hence we can learn from him and, if need be, we can engage with him in intelligent dialogues rather than in shouting matches. He is, in sum, a *pre-postmodernist* scholar, as Merton might have called him.

Mario Bunge, FRSC
Department of Philosophy
McGill University, Montréal

Introduction

Controversies at the Crossroads of two Specialties

Scientific controversies, which may provisionally be described as "organized debates" whose aim is to establish or verify our knowledge of the world, fall within the competencies of two specialties in the field of sociology.

If one places the emphasis on *form*, controversies may be compared to the public debates or oratorical jousting that fall into the category of *conflicts*. Debates, however, must be distinguished from other forms of conflict. As Rapoport (1974) writes: "It is clear that here neither the *harming* of the opponent nor the '*outwitting*' of the opponent is relevant to the objective. The objective is to *convince* your opponent, to make him see things as you see them [...] If this objective is kept constantly in mind, the debate appears to be a process very different from either the fight or the game, and different concepts must enter the analysis of this process" (11).

If instead one focuses on *content*, then controversies are conflicts that have to do with declarations tied to a specific activity, that is, to the production of verified knowledge regarding the world. In other words, *scientific* controversies may develop in specific ways and assume sets of rules that do not correspond to those of political, religious, or artistic controversies, and this is precisely because of the specificity of the knowledge that is being brought into question in this type of debate. And it is for this reason that the study of scientific controversies must take into account advances in the *sociology of scientific knowledge*— henceforth referred to as SSK.

Below I will describe the general orientation and principal results of studies conducted in these two areas.

1. The Sociology of Science

The sociology of science is a specialty stemming from the sociology of knowledge, and in particular the *Wissensoziologie* of Max Scheler

1

(1926) and Karl Mannheim (1952). Formulated at the end of the 1930s by Robert K. Merton, for many decades the sociology of science continued to reflect the original impulse given to it by its founder.

In the wake of his doctoral thesis on advances in the experimental sciences in seventeenth-century England, Merton began to elaborate a series of ideas on the nature of scientific activity and its various aspects. To begin with, Merton was responsible for an important breakthrough in the methodology of research in this field. In an article first published in *Isis* (1937), he set out five paradigms for investigations in the sociology of knowledge (1973, 7–40). When confronted with a specific item of knowledge, the sociologist should ask himself the following questions:

1. *"Where is the existential basis of mental production located?"* What is known of the actors who adhere to a given knowledge set, be it revolutionary or traditional, from a social (class, generation, occupational role, etc.) or a cultural perspective (values, ethos, *Weltanschauungen, etc.*)?
2. *"What mental productions are being sociologically analyzed?"* What is being submitted for analysis, that is, what spheres (scientific theories, political ideologies, moral or religious beliefs, etc.) and what aspects (foci of attention, levels of abstraction, the theory-practice relationship, the experimental content, etc.) of knowledge are being engaged?
3. *"How are mental productions related to the existential basis?"* Are we dealing with a causal-functional link (determination, conditioning, functional interdependence, interaction, etc.) or a semantic link (compatibility, harmony, congruence, analogy, etc.)?
4. *"Why [are they] related?" Both manifest and latent functions can be imputed to these existentially conditioned mental products.* To what end are mental products linked to their existential basis? Is the purpose to produce useful knowledge, to control nature, to orient one's actions, or to reinforce one's power?
5. *"When do the imputed relations of the existential base and knowledge obtain [i.e., manifest]?"* Under what sociohistorical conditions may a mental product appear? Are we dealing with knowledge that is universalizing in its impact or confined to particular societies or cultures?

These five questions delimit a canonical approach to issues in the sociology of knowledge. But it may be noted as well that points 4 and 5 contain in embryonic form all the elements that set SSK apart from and render it independent of the sociology of knowledge. In effect, the manifest functions of a scientific theory do not correspond to those of a political ideology, whose objective is to galvanize the loyalty of supporters, attain power, etc. (point 4). Furthermore, contrary to religious dogmas, scientific concepts are inherently universalizing forms of knowledge (point 5). Having arrived at this conclusion, Merton rightly examined the question of the autonomy of scientific knowledge and above all what guarantees its autonomy within scientific institutions and within society itself.

The *Ethos of Science* first published in 1942, seeks to address these questions, and to define the moral principles that ought to govern the activity of scientists. According to Merton: "The ethos of science is that affectively toned complex of values and norms which is held to be binding on the man of science" (1973, 267–278). On the understanding that its norms are not technical but deontological in nature (unlike, for example, the MKS system of units), Merton's concept can be summarized in terms of some of its principal traits: universalism, communalism, disinterestedness, and organized skepticism. Numerous criticisms against Merton's ethos of science have been raised (see Barnes 1970). However, many critics base their arguments on the search for counterexamples, an approach that is destined to founder, not because the anomalies that they underline have no basis in reality—Merton himself recognized their existence (1973, 274–276, 309–316)—but because the link between these counterexamples and the norms of the scientific ethos remains insufficiently elucidated. The link does not appear to be causal and therefore is not consequential. In the sciences no behavioral norm can be refuted simply on the basis of incidents of deviant behavior. It is invalidated only if such behavior succeeds in winning the approval of the scientific community or, at the very least, does not lead to severe reprobation or official sanctions. In fact, when they are discovered, the perpetrators of fraudulent research generally pay a heavy price in terms of their professional reputations and the credit that they are accorded for their work.[1] To my knowledge, there is no example of a researcher who has trespassed the norms of the scientific ethos and been praised by his colleagues. On this point, Merton's conceptualization remains entirely pertinent.

One important particularity of Merton's work is the attention that he drew simultaneously to the principles governing the organization of scientific activity and the sometimes unexpected consequences stemming from their application. Thus "peer review"—one of the pillars of the scientific community's self-policing and validation process—is meant to ensure that there is a degree of "joint-ownership" in the different tasks involved in scientific research (for example, a referee is responsible for the decision regarding the articles that pass across the editor's desk, but it is not his *profession* to verify the scientific validity of papers being submitted for review).

However, this apparently equitable system translates in the real world into an unassailable *gerontocracy*. Referees are chosen because they are experts in a given area, and a specialist is recognized on the basis of the number and quality of the papers he has published. However, to conduct research of a high standard requires time and, as a result, young scientists are rarely called upon to serve as referees (Merton 1973, 537–545).

In another example of a perverse effect, the supposedly meritocratic system in which a researcher is rewarded for his or her discoveries is rendered nonegalitarian by virtue of the *Matthew effect* that can be summed up as: "The rich get richer and the poor get poorer" (Merton 1973, 439–459). Extending this to the realm of science, a researcher who already has a well-established reputation enjoys a much greater chance of winning further recognition. If an article of multiple-authorship is given to a group of physicists to read, and they are then asked to name the authors, they will frequently remember only the most prominent and not be able to recall the (younger or more obscure) coauthors. Prestige attracts further prestige.

Somewhat paradoxically—since the sociology of science is an offshoot of the sociology of knowledge—the Mertonian school was interested above all in the study of the social factors that orient the *production* of knowledge, rather than the factors that orient the *knowledge* itself. This trait can be deduced from the fact that, according to Merton, scientific knowledge represented universalizing knowledge that is detached from the sociohistoric conditions out of which it arose. Thus, the sailor has no reason to remember that the compass he is using was invented by the Chinese, and an accountant balancing his books has no need to recall that the notion of zero originated in ancient India. The universal conditions that govern the use of the compass and the zero render obsolete—except perhaps for the historian who is studying the conditions for the production of scientific knowledge—the link between these inventions and the social context underlying their discovery. As a consequence, until quite recently sociologists tended to leave the analysis of scientific knowledge to epistemologists and science historians.

The publication of *The Structure of Scientific Revolutions* by Kuhn (1962), allied to the rediscovery of certain fundamental texts such as those of Duhem (1914), Fleck (1935), and Polanyi (1958), marked the rise of an entirely new way of envisaging the sciences that has come to be designated the *social turn.* The most prominent feature of this sociological current—which broke with the tenets of the Mertonian school—is that scientific knowledge can in and of itself constitute an object of sociological study. Thus the idea of the SSK was born, both as an extension of the sociology of science and as an attack on the traditional territories of epistemology and the history of science. Often detectable in the work of this new generation of sociologists was the desire to reestablish ties with the earliest (pre-Mertonian) sociologists, such as Durkheim and Mannheim. Particular attention was focused on the theme of scientific knowledge as a social construct, and no doubt this is why scientific controversies became a specific subject of study within the specialty.

Without pretending to offer an exhaustive list, at least fifty publications could be mentioned that explore the question.[2] For a variety of reasons, most of the scholars who have undertaken the study of scientific controversies are firm adherents to the doctrine of relativism (cf. chapter 1). Their texts contain clear indices of the importance ascribed to the subject of scientific controversies. For instance, in the introduction to their anthology *La science telle qu'elle se fait*, Latour and Callon refer to *"une entrée royale par les controverses"* (1991, 26). It was only later that criticisms began to be directed against this approach.

2. The Sociology of Conflict

The sociology of conflict long predates the sociology of the sciences, and as a specialty has given rise to a vast literature, among which the work of Simmel (1908), Coser (1956), Rapoport (1960) and Caplow (1968) are particularly salient.[3]

Georg Simmel's study of conflict (*Streit*) stands apart because of the author's somewhat idiosyncratic presentation of his material and the extremely wide range of cases studied. Simmel (1964) underlines the fact that, despite their antagonism, the parties to a conflict "fight under the mutually recognized control of norms and rules [. . .] This same phenomenon is characteristic, finally, of all conflicts in which both parties have objective interests" (35, 38). The second important proposition of Simmel is that *social factors* may be responsible for the intensity of an antagonism, for instance, the existence of shared characteristics or the affiliation to a shared context (43). Close ties may give rise to a more vehement antagonism; thus, the degree of tension in a debate between researchers belonging to the same scientific community and working in the same field will probably be higher than is considered usual. According to Simmel, the intervention of a mediator can reduce the level of antagonism by expurgating the conflict of its emotional and aggressive elements. Finally, the author's discussion of different forms of conflict resolution (111) deserves attention, not so much for the precision with which he lays out his empirical typology but rather for certain percipient remarks arising out of his current work on the resolution of scientific controversies. While it is difficult to pass a global judgment on Simmel's study of conflict, it may be said that those of his observations which touch directly on the area of science[4] appear to be less important than his general sociological propositions, although the conditions required for the transposition of these propositions to scientific controversies remain to be explored.

Lewis Coser's study, *The Functions of Social Conflict* (1956; 1982), expands on the observations of Simmel but adopts a more systematic approach. The most interesting aspect of Coser's analysis is his attempt to rectify a lacuna in the functionalist tradition, which has always turned away from an analysis of conflict. Merton (1953) had already criticized this shortfall (96–99), but it was Coser (1982) who first undertook a functionalist analysis of conflict. Three propositions in his book demand attention. *Primo*, the more closely linked two adversaries are, the more violent the conflict will be, whereas a strong mutual dependency between the parties or individuals involved will prevent the development of social cleavages (53).[5] *Secundo*, a conflict between two groups will reinforce the sense of cohesion within each group (63). *Tertio*, a conflict will be resolved all the more quickly if it takes place within an institutional context (104). Coser's results are convincing, but their extrapolation to controversies in the realm of science cannot be taken for granted, because it is difficult to assert that the interests which come into play in a scientific debate are the same as those found in a social conflict.

In his research on coalitions, Theodor Caplow (1968) subscribed to Simmel's theories on conflict. He begins with an examination of the role of third parties in a conflict. This third party may be either a *mediator*, a *tertius gaudens*, or an *oppressor*. In the first case, the introduction of an impartial *mediator* presupposes the absence of a coalition between the interested parties. In the second case, a *tertius gaudens* exploits to his advantage the dissensus between rival parties. In the third case, an *oppressor* seeks to foment conflict between two parties to serve his own interests. On this basis Caplow studies the types of coalitions that may form within a triad.

After describing the results of an experiment using the Indian board game of *pachisi*—which pits four players divided into two teams against each other—to analyze behavior and outcomes, Caplow passes in review a number of case studies that allowed him to study the formation of coalitions. For example, a detailed analysis is presented of the dispute that arose in 1949 between the Chicago Housing Department, the city council, and the mayor on the construction of new public housing (147). The view of the Housing Department was that it was necessary to increase the amount of residential housing in the city, while the city council thought that it would be better to pursue a broad program of urban renewal. The (politically liberal) Housing Department attempted to form a coalition with the mayor in order to force the (conservative) city council to adopt their proposal. Initially the mayor tried to maintain a neutral position, but in the end he threw his support behind the city council. Given his privileged position, during the course of the dispute the mayor became the object of differing expectations. However, as Caplow underlines, "The Mayor, like the other participants, was moved by the forces of the situation, but he enjoyed enough freedom of action to tip the balance, or seem to do so, at the turning points of the conflict".

I have described this case in some detail because it illustrates perfectly the difficulties that one may encounter when applying the findings from the sociology of conflict to scientific controversies. Can the type of political coalition that has just been described—in which a disagreement is decided in favor of the most powerful party—be transposed to scientific debates? The question remains open. It is easy to see that two lines of argument could be followed in the search for an answer. Arguing from the generality of human behavior, some might be tempted to respond in the affirmative. Others, arguing from the specificity of human conduct, would respond in the negative. And here in fact lies the difficulty: one may intuitively sense that findings from the sociology of conflict can be applied to scientific debates, but there is no way of knowing exactly to what degree. The literature on scientific controversies runs the entire gamut from a maximalist position (in which the social and scientific interests are comparable) to a minimalist position (the social and scientific interests are fundamentally different). In effect, in the case of the minimalist position the sphere of interests and values is not homogenous in structure; it is rather a specific construction—erected by the social

7

actors—of the interests and values pertaining to their specific activity. This specificity means that one can no longer place on an equal footing such differing forms of conflict as a private dispute, a commercial rivalry, a political debate, and a scientific controversy. The rules of the game each time will be completely different.

3. Elements for a Classification

It would be useful at this point to provide a preliminary definition of controversies, if only to limit our use of the term to comparable phenomena. From the sociological point of view, a *scientific controversy* is characterized by: *the persistent and public division of members of a scientific community (either individually or as coalitions), who sustain contradictory arguments in the interpretation of reality.*

This is why one might say that, in its very constitution, every controversy involves an act of *separation* (Raynaud, 1990) and presupposes the involvement of at least two actors. In the most elementary of cases, one may be dealing with simple "individuals-persons." All the same, controversies will frequently draw in "macroindividuals"composed of numerous researchers at the head of whom one will generally find leaders. In this definition, the notion of a "persistent and public division" is a direct echo of the work of McMullin (1987). The reference to a "scientific community," which confers a sociological dimension on the object of study, has been adopted in line with the recommendations of Engelhardt and Caplan (1987). As these authors observed, it is often difficult to come to a unanimous conclusion in a controversy because the debate has spread into the most heterogeneous of *milieux*.[6] The phrase "contradictory arguments in the interpretation of reality" underlines the fact that a controversy is not the same thing as a "dispute over priority," such as the quarrel that sprang up between Newton and Leibniz over the invention of differential calculus, or the dispute between Luc Montagnier and Robert Gallo regarding the discovery of the HIV virus. A disagreement over priority does illustrate a *separation* between two researchers or two research groups, but in this case both parties accept the *same interpretation* of the phenomenon under study.

Scientific controversies can be classified and analyzed within the framework of eight characteristics, which are presented here.

1. The Object

McMullin (1987) has proposed that, in a scientific debate, distinctions can be made based on the object under dispute. It is necessary to differentiate between controversies over facts, controversies over principles (be they methodological or ontological), and controversies over theory.

One would place in the first category the "Battle over the Electron" which, beginning in 1910, opposed Robert A. Millikan and Felix Ehrenhaft in a contest to determine the charge of the electron.[7] The two scientists, one in America and one in Austria, began their experiments using different equipment and approaches, but their methods gradually converged without their ever arriving at an accord. Moreover, even though the stakes were of the highest for the science of physics, the theoretical work being conducted at the time had no direct influence on the measurements. The controversy was based essentially on the facts.

In the second category one may cite the controversy over the Hubble constant H_0, which since the 1970s has seen the supporters of the "long scale" of Sandage and Tammann lined up against the advocates of Vaucouleurs's "short scale" with regard to the rate of expansion of the universe. In this case the discordant results reported can be ascribed to the methodologies used by the different research teams.[8]

The "Guerres perspectives" of the 1640s provide an illustration of the third category of controversies. The principal actors in this war were Girard Desargues and Jean Dubreuil,[9] and the source of the controversy lay in the fact that Desargues had introduced a new conceptualization of perspective—projective geometry, whose keystone was the theorem that bears his name[10]—a formulation that Dubreuil, who was still wedded to the classical theory of perspective, dismissed as useless. However much the practical utility of perspective and projective geometry, only the theoretical component was under contention in this seventeenth-century controversy.

The categories laid out by McMullin have only an *indirect* sociological dimension. The distinctions between the objects of a scientific controversy are in fact of little interest to the sociologist except in those cases where the debate—be it over the facts, the method, or the theory—leads to changes in social behavior. One might surmise that controversies over a theory or principle would give rise to more prolonged argument and draw in more participants than controversies over facts (see *infra*: the discussions of *Extent* and *Duration*). Indeed, a factual statement being closer to the experimental test than a theoretical statement, it is more easily assessable. As such, there is a greater chance that a factual declaration can be quickly corroborated or disproved. In these cases the adversaries will eventually agree on one of the initial hypotheses—on condition that the contested point can be subjected to a test devoid of ambiguity and that the two parties judge this test to be sufficiently discriminating.

2. Polarity

Polarity refers to the number of opposing camps that face each other during the course of a scientific debate. The most frequent case is that of the *bipolar* controversy conducted by two parties adhering to diametrically opposed positions—for example, physicists at the end of the nineteenth century who had to declare themselves for or against the existence of ether. Nonetheless, there exists the possibility of disputes involving more than two poles, such as *tripolar* controversies. The internal structure of the problem[11] and specific sociological characteristics would appear to lie at the origin of multipolar disputes. It may be conjectured, for example, that the interpretations of a phenomenon will be all the more nuanced the larger the number of researchers making up the scientific community, with a concomitant multiplication of the number of interpersonal ties.[12] However, the possibility cannot be excluded that the bipolarity so often observed in controversies may itself have a sociological basis. In his analysis of social conflicts, Coser (1956; 1982) suggested that the internal cohesion of a group is reinforced by conflicts with outsiders. If during a period of dissensus one considers each camp in a controversy to constitute a group—or at least a "latent group"—this proposition becomes perfectly intelligible.

Coser (1982) writes: "A situation of conflict binds members to one another as closely as necessary and compels them to act in as uniform a manner as is required, whether they are marching forward in unison or unilaterally repulsing an attack" (63). This is perhaps why, as a controversy intensifies, it tends to assume a bipolar configuration. Such polarization is often followed by an equally remarkable development: the passage from the configuration "A or B" ("black or white") to the configuration "A or not A" ("black or not black"). This transformation can be explained by the fact that only a set of alternatives expressed in the form of an affirmation–negation ("A or not A") is productive of knowledge. And this in turn is because, as long as an experiment can be designed to test them, at least one of the two theses will be refuted. In contrast, a test applied to a controversy of the type "A or B" may produce results that do not support either of the two hypotheses.

3. Extent

Taken at time *t*, a controversy may be more or less generalized depending on whether it originated in a disagreement between isolated scientists (for example, al-Bīrūnī and "a great savant"—probably Ibn Sīnā—regarding the rotation of the earth on its axis) or if large numbers of scientists joined the fray at the outset. It is well known that the debate between William Harvey and René Descartes regarding the circulation of the blood was taken up by an extended network of friends and disciples long before the establishment of the first learned societies and academies of science in England and France (Mendelsohn 1987). Scientific controversies can therefore be either *narrow* or *broad* in extent. What is more, it is not rare to find them springing up between researchers with allegiances to different institutions, coming from different professional backgrounds (Ben-David 1991), or committed to different research methods (Roll-Hansen 1980).

Lewis Coser's work (1982) indicates paths to investigate the social factors that could contribute to the extent of a conflict. First of all, he proposes that one distinguish between two elements—the size of the group and the degree of involvement of its members—because there can be a negative correlation between these two variables.[13] The more restricted the circle, the higher the degree of engagement of its members (otherwise, their efforts would be insufficient to make their voices heard). In contrast, the larger the group the smaller the commitment demanded of each member. This size effect can have direct consequences on the scientific content discussed during the controversy. A small group will always require a more unified consensus, because "any internal dissent will place at risk the mobilization of the energies necessary to face the external conflict" (73).

This reasoning provides support along a different line for the remarks made above regarding the bipolarization of controversies, suggesting the outlines of a response to the observation that during a controversy involving a large portion of the scientific community, "leaders" tend to emerge in each camp (such as Needham and Spallanzani in the earliest debate on the theory of spontaneous generation). This phenomenon should not always be interpreted as a creation of hierarchies within the camps engaged in the controversy. It rather illustrates the distinction between active and passive agents in a controversy, the latter being satisfied to remain in the background and produce results to corroborate the thesis being actively defended by the leader of the group.

4. Intensity

As in all debates, a scientific controversy can produce more or less violent exchanges between the contradictors. Certain disagreements are resolved without causing a ripple in the community, whereas others may bring the participants to the point of paroxysm. The virulence of Felix Pouchet's opposition to Louis Pasteur and the infamous accusations of Dubreuil against Desargues contrast markedly with the instances of courteous consideration recorded by Merton as genuine cases of "noblesse oblige." More than one example of the latter type of behavior has been documented. It is known that Leonhard Euler and Joseph Louis Lagrange were both working on a "calculus of variations" and that Euler delayed publishing his results until his colleague had completed his calculations. Judging Lagrange's solution to be superior to his own, Euler recommended his nomination as a corresponding member of the Academy of Sciences in Berlin. In the same vein, Hooke writes to Newton: "Your design and mine are, I suppose, both at the same thing, which is the discovery of truth, and I suppose we can both endure to hear objections, so as they come not in a manner of open hostility" (Robert Hooke to Isaac Newton, February 1675/6). Scientific factors (the nature of the knowledge in question), psychological considerations (the tenacity and fierceness of the rivals), and other elements can help to explain the intensity of a particular controversy. Thus it appears that the furious levels reached in the debate between Pasteur and Pouchet in the 1860s can be ascribed to emotional more than to scientific causes. However, for a controversy to reach a minimum level of intensity, it is necessary that each side acknowledge the existence of the other (see *infra*: type of recognition). Without this there can be no escalation in the debate.[14]

There are sociological factors underlying such differences in intensity. Returning to the notion of Simmel that the narrower and more exclusive the relationship between adversaries, the more violent the conflict, Coser suggests that: ". . . the interdependence of groups and individuals in modern society prevent, up to a certain point, any fundamental cleavages" (Coser 1982, 53). The greater the degree to which an individual is integrated in a diversified range of social circles, the lower the risk of an exact superimposition of the causes of microconflicts between these segments.[15] Inversely, the more strongly an individual associates himself with a single closed group, the higher the risk that internal conflicts will mutually reinforce each other and split the group into intractable camps. The intensity of the conflict with an external party will therefore be directly proportional to the internal homogeneity of the group.

What is more, in the case of segmented participation, the microconflicts that can arise between various segments will prevent an individual from investing all of his energy in a single sphere of conflict (Coser 1982, 55). This notion would allow us explain why certain controversies are resolved quietly, while others seem to escalate and attain a high threshold of visibility. Above all this illustrates that there is perhaps more a difference of *degree* than of *nature* between science in the midst of a crisis and what is deemed the normal, everyday conduct of science.

5. Duration

Some controversies may be *limited in duration*, but the time frame of those touching on problems which are complex in nature or not entirely susceptible to verification by experiment may be quite *long*, perpetuating themselves through successive generations of researchers. The dispute begun by Priestley and Lavoisier over the phlogiston theory and the nature of combustion lasted from 1770 to 1796. In the field of optics, the opposing theories of extramission and intromission which had been posited in antiquity still formed a topic of lively discussion in the sixteenth century. Apart from a handful of cases of extreme simplicity, the duration of a dispute will often depend on the perspective adopted in the definition of the problem.

It may in fact happen that the terms of a problem will be more or less radically redrafted from one generation to the next. For example, in the seventeenth century the issue of spontaneous generation revolved around the procreation of insects, but by the nineteenth century the question only concerned microorganisms. The sphere remained biology, but the scale of the phenomenon was different, the methods of observation had altered, and the problem being posed was no longer the same. Depending on the perspective adopted, it could be said with equal correctness that the controversy extended from the seventeenth to the nineteenth centuries, or that it took place during the first half of the nineteenth century, or that the argument only applied in *stricto sensu* to the exchange of views between Pasteur and Pouchet and therefore lasted five years, from 1859 to 1864.

It appears that the institutionalization of a controversy, which implies prior agreement on the modalities for verifying the truth of a hypothesis and the conditions that must be met before victory is declared by one side or the other, will limit the duration of a conflict. The modalities and signs of success also serve to clarify the public perception of the outcome of a polemic. As Lewis Coser writes: "In conflicts that are not fully institutionalized, [. . .] the loser is not obliged to publicly acknowledge that he has lost, or even to admit it to himself" (1982, 105). The expert committees created by various academies of sciences in the nineteenth century offer exemplary instances of the institutionalization of scientific conflict. They were convened to adjudicate on affairs in which the opposing camps had presented their arguments in various forums without managing to arrive at an agreement. Experiments conducted in front of independent experts allowed the contested point to be settled.

6. Forum

Collins and Pinch (1991) adopt the classic distinction between the types of *forums* available for a debate, that is, the resources and authorities used by the opposing sides to advance their point of view. One may distinguish between the *constitutive forum* (conducted between peers and based on the formulation of theories, experimentation, publication in scholarly journals, presentations at meetings, etc.) and the *contingent forum* (conducted in the public sphere through popularizing articles, advertising, direct appeals to public opinion through various media, etc.). In order for actions to take place in the *constitutive forum*, they must be "founded on universalizable, non-contingent premises" (303).

All scientific controversies unfold in the *constitutive forum*, but it is rare that a dispute of any importance will be limited to this arena. One can presume that the debate will spread into the *contingent forum* with a speed that reflects the degree to which it is felt to resonate—whether justifiably or not—with certain social problems or specific aspects of issues currently under discussion. In this way, the path for its introduction may have already been opened by the public's familiarity with a related issue. Such was the case with the theory of spontaneous generation, which in a way drew support from the coeval debate on evolutionism versus creationism. It is the case today with the controversy over climate change and whether the Earth's rising temperature should be ascribed to solar phenomena or anthropic factors. That the work of Svensmark[16] met with such violent opposition in climatology circles can be explained first and foremost by the fact that the greater part of the research on global warming is filtered and relayed by the IPCC (*Intergovernmental Panel on Climate Change*), which counts some 3,000 researchers around the world. However, the task of the IPCC is not only to communicate scientific information; its mission is also to provide political decision-makers with expertise by evaluating the impact of climate change and presenting proposals for appropriate strategies to counter the threat. This commitment to action, which is reflected in the many reports published by the IPCC (2001, 2013), is responsible for the reluctance of these experts-*cum*-advisors to entertain any new and conflicting theory that could ruin the progress that has been painstakingly made over decades. This is why the hypothesis that greenhouse gases are the cause of global warming remains the official thesis of the IPCC. The "relay effect" ensured by the organization, which in this case is acting as a *contingent forum*, continues to influence scientific debate on the question.

7. Recognition

One can draw a distinction of principle among controversies based on the number of parties that actually recognize the existence of a dispute. In the simplest case involving two opposing views, if the acknowledgement of a divergence emanates from both camps, then the controversy is *bilateral*; if instead it emanates from only one camp, then the controversy is *unilateral*. The dispute over spontaneous generation was a bilateral controversy, as is the one that divides cosmologists today regarding the determination of the Hubble constant H_0, with values ranging from 45 to 95 km/s/Mpc. Both Sandage and Vaucouleurs recognize the existence of this disagreement. The controversy between the medical schools of Paris and Montpellier was instead of the unilateral type; the division between the vitalists and the organicists was more often proclaimed by the disciples of Paul Joseph Barthez at Montpellier, who were seeking to maintain the past glories of their faculty, than by the Parisians, who remained secure in their experimental approach. The latter therefore continued unperturbed to develop their methods without paying much attention to (or at least feigning to ignore) the objections regularly raised by their colleagues in Montpellier. In the eyes of the Parisians, the controversy did not even exist.

8. Settlement

Collins and Pinch (1991) distinguish between the ways in which certain initial hypotheses are finally judged to be inadequate. They refer to *explicit rejection* when the hypothesis advanced can be subjected to experimental testing or to evaluation by experts or to comparison with previously published reports, and *implicit rejection* when the hypothesis advanced is received with unanimous incredulity (304). The ideas of Pouchet regarding spontaneous generation were explicitly rejected and this consensus was duly formalized by a *commission d'expertise* set up by the French Academy of Sciences.

However, even in the case of issues judged by all to be worthy of attention and that are therefore debated in the *constitutive* forum, the settlement of the controversy may not always assume an explicit form. For example, research on the elementary charge of the electron, described as "the most fundamental question" in physics in the 1910s, did not conclude with an outright repudiation of the work of Ehrenhaft. As Holton (1982) writes, his results were simply allowed to "fade into obscurity [. . .] Indeed, Ehrenhaft continued to publish on subelectrons into the 1940s, long after everyone else had lost interest in the matter" (79).

It appears that the kind of settlement that can be reached is linked both to the nature of the knowledge in question and to the type of acknowledgement that is made of the controversy. In the case of a unilateral controversy, there is little likelihood that the group of researchers being challenged will take the trouble to refute the arguments of adversaries whom they are pretending to ignore. This implicit rejection can also be explained by the high cost of entering into the debate that recognizing the criticism would involve. In order for researcher B to agree to refute the theses of researcher A, the benefit expected must be superior to the cost of engaging in the quarrel. If A is of a lesser status or credibility than B, then B will refuse to enter into the dispute because the effort and measures required would be disproportionate to the advantage gained. Philip Kitcher (1992) formulates a model of this phenomenon (248). In the case of an implicit rejection, the amount and quality of the scientific work conducted by those adhering to one or the other school of thought could be a determining criterion in the settlement.

To use the terminology of Mendelsohn (1987) and McMullin (1987), one must differentiate between the *resolution* of a controversy, which implies the discovery of a rational solution; the *closure* of a dispute, which implies a formal procedure for settling a controversy, even if this does not necessarily result in an accord between the parties; and the *abandonment* of the argument, in which there is a simple cessation of discussion of the matter. More subtle distinctions have been proposed, such as those of Engelhardt and Caplan (1987):

1) *Closure through loss of interest*: all parties may simply tire of the discussion, and the disagreement will evaporate of its own accord.
2) *Closure through force*: certain agents may mobilize outside resources, such as government authority.
3) *Closure through consensus*: the agents arrive at a nonscientific interpretation that translates into the adoption of a belief.
4) *Closure through sound argument*: the agents reach an agreement based on generally accepted scientific criteria, in *lato sensu* if the inference and the proof are influenced by historical or cultural factors, and in *stricto sensu* if they are not.
5) *Closure through negotiation*: when the process by which the agents arrive at a settlement does not correspond to any of the first four modalities. In this case, the parties will continue the debate after agreeing on a set of procedural rules that will probably allow them to arrive at a conclusion.

It may be noted that modes 1, 2 and 3 in this classification system do not lead to the rational resolution of a controversy based on the facts, whereas modes 4 and 5 do, the first in a natural process and the other in a somewhat forced manner. Therefore, some of Engelhardt and Caplan's categories have been criticized because of their insufficient descriptiveness in the case of scientific controversies (Macklin 1987, 620).

The eight characteristics that have been described above—*object, polarity, extent, intensity, duration, forum, recognition,* and *settlement*— could be helpful in delineating the distinctive characteristics of the scientific controversy. Some of them may also be used to describe conflicts taking place outside the field of science, whereas others seem to be specific to debates on scientific topics. These eight traits in a certain sense lie at the origin of the study of the cases presented in this book.

The first chapter proposes a discussion of the relativist and rationalist arguments that lie at the heart of the study of controversies. The essays that follow will undertake to evaluate these arguments not through a critique of their underlying principles, but based on the study of relevant cases from the domain of SSK.

The structure of this book is quite simple: every chapter will examine a separate question that is central to our theme. The principle of cumulative asymmetries, the social determination of scientific knowledge, and the historicity of the norms of rationality will each form the subject of a specific study that will take as its starting point a different controversy in the history of science. Different in terms of their sociohistoric context: medieval Islam and the Chaghatai khanate in Transoxiana, England in the thirteenth century, Europe in the nineteenth century, and so on. Different in terms of the disciplines approached—from mathematics to optics and from medicine to microbiology. Different, finally, in terms of their inherent characteristics: their extent, the type of forum in which the argument was conducted, the type of recognition underlying their resolution, and so forth. The controversies selected for discussion here were pivotal to the history of the sciences and could help us to understand the origins and evolution of the some of the more recent scientific debates.

Notes

1. Certain cases of fraud won great notoriety. In 1987 Paul Chu submitted a manuscript to *Physical Review Letters* describing the composition of the first commercially viable superconducting compound, but his formula was partially falsified because he substituted one component (Ytterbium, abbreviated Yb) for another (Yttrium [Y]). The German Research Foundation (*Deutsche Forschungsgemeinschaft*) verified that out of 347 papers published by Friedhelm Herrmann, ninety-four contained falsified data. In 1991 the Australian scientist John Talent unmasked the Indian paleontologist Viswa Jit Gupta, nearly all of whose publications were based on fabricated fossils. Gupta's offer to resign his position at Panjab University was accepted by the university's senate. However, he wrote a second letter withdrawing his resignation, and the court allowed him to return, causing great consternation within the academic community. Between 1976 and 1979 John Long published falsified results of cell culture studies on Hodgkin's disease that were then cited in many research manuals. In 1981 his deception was uncovered by the journal *Science* and the author was forced to acknowledge that he had manipulated his data. The news created fury in the scientific community and all references to Long's work were eliminated from textbooks and manuals. Long himself experienced great difficulty in finding colleagues willing to collaborate with him afterwards, or to receive credit for any research that he did.
2. Schmitt (1967), Farley (1972), Bechler (1974), Burnham (1974), Farrall (1975), Harwood (1976), Frankel (1976), Roll-Hansen (1979), Roll-Hansen (1980), Cantor and Hodge (1981), Callon (1981), Lemaine and Matalon (1985), Stuewer (1985), Rudwick (1985), Engelhardt and Caplan (1987), Gálvez (1988), Latour (1989), Scott, Richards and Martin (1990), Farley and Geison (1991), Shapin (1991), Callon and Latour (1991), Shapin & Schaffer (1993), Collins and Pinch (1994), Kim (1994), Fabiani (1997), Schweber (1997), Machamer, Pera and Baltas (2000), Kurrer (2008), Tahiri (2008), Vandenbosch & Vandenbosch (2008), Zemplén (2008), Moore and Stilgoe (2009), Xu and Cheng (2009), De Freitas and Pietrobon (2010), Venturini

(2010), Bechler (2012), Goodwin (2013), Junges (2013). There is also a lot of research on controversies in the fields of argumentation theory (Dascal 1998; 2000; Barlotta and Dascal 2005; Dascal, Racionero & Cardoso 2006; Zemplén 2008; Demeter and Zemplén 2010; Li 2010, 2011), communication (Brossard 2009; Besel 2011) and psychology (Jensen 2012; also tangentially Barlotta and Dascal 2005).

3. I will leave aside Max Weber's reflections on the concept of "struggle" (*Kampf*) (1995, 74–78), because he limited himself to fairly broad definitions regarding the subject under consideration here. According to Weber, controversies should be placed in the category of "peaceful contests," that is conflicts that do not involve the mobilization of physical violence. They take the form of competitions governed by a set of rules, which he termed *geregelte Konkurrenz*. In addition, the German sociologist emphasized the process of selection at work between the competitors based on the characteristics innate to the type of confrontation envisaged. However, with regard to this point he does not directly address the case of the scientific debate.

4. That is to say: "The contrast between unity and antagonism is perhaps most visible where both parties really pursue an identical aim—such as the exploration of a scientific truth" (1964, 39). "Similarly, ambitious competition in the field of science aims not only at the adversary but at the common aim, on the assumption that the knowledge gained by the victor also is the gain and advantage of the loser" (1964, 59). "[Competition] uses the production of objective values as means for attaining subjective satisfactions" (1964, 60).

5. In a study combining sociology and anthropology, Flap (1997) rediscovered this result without relying on the demonstrations of Coser, thus providing verification of the robustness of Coser's proposition. Beginning with an analysis, conducted in utilitarian terms, of the application of social networks, Flap shows that "conflicts are less violent when the social circles overlap" (183).

6. Engelhardt and Caplan write: "One is brought to dissecting what appears at first sight to be one controversy into a number of controversies, where 'a' scientific controversy is defined by the existence of 'a' community of disputants who share common rules of evidence and of reasoning with evidence. If such rules are not shared, then the dispute is not a single scientific controversy [...] Each community defines the controversy in a different fashion, and initially, at least, they do not share means for resolving the controversy" (1987, 12).

7. Millikan and his collaborators set up an apparatus consisting of an expansion chamber containing condenser plates that could be electrically charged. Fine particles blown by an atomizer into the chamber would therefore be subject to two forces, the force of gravitation and that imposed by the electric field. They could study the movement and velocity of the falling particles and deduce their electric charge. Millikan first used droplets of water, but these were eventually replaced with drops of oil to avoid the effects of evaporation.

Ehrenhaft constructed a very similar apparatus, but carried out his measurements on particles of metal (platinum and silver) produced by an electric arc. Furthermore, instead of making a single series of measurements on particles subjected to both the force of gravity and an electric field, he always conducted two sets of measurements—one on particles falling in a gravitational field alone, and one on particles exposed to the forces of the gravitational and electric fields simultaneously.

Millikan found all of the charges he measured to be integer multiples of an elementary charge $e = 4.774 \pm 0.009 \times 10^{-10}$ u.e.s. (data published in *Physical Review* in August 1913). During this same period Ehrenhaft announced several times that he had measured electrical charges much lower that that of the electron; in this way the sub-electron entered the scene. However, the values recorded by him were so disparate—between 1.38×10^{-10} and 7.53×10^{-10} u.e.s.—that in the end Ehrenhaft renounced his atomist theory of the *elektrische Elementarquantum*. The Austrian scientist's personality and questions regarding the selective publication of data were sensitive points in this controversy, so much so that Lorentz in 1916 and Chwolson in 1927 judged the debate over the electron charge to have not yet been definitively resolved (Holton 1982, 133–232; Boudon 1990, 151–154).

8. The terms "long scale" and "short scale" derive from estimates of the age of the universe based on the Hubble constant. In the first case, according to Sandage the universe was some 19 billion years old while in the second, by Vaucouleurs's estimate it was a mere 9.5 billion years old. Beginning with the measurements effectuated by Hubble and Humason in 1929 ($H_0 = 500$ km/s/Mpc), the proposed value of the Hubble constant has varied considerably. As Leverington observed: "Today (1994) the value of H_0 is still unclear, with estimates ranging from 45 to 95 km/s/Mpc" (1995, 237). Hubble's law, $V = H_0 r$, states that the recession velocity of the galaxies (deduced from the observation of their redshift) will vary in direct proportion to their distance from the measurement point. The principal difficulties concern the measurement of recession velocities and, above all, the determination of the distance of the galaxies: 1) The more distant the objects, the more tenuous their spectral trace will be (although some progress has been made using the HST faint object spectrograph). 2) Some argue that the redshift is exclusively the result of the radial movement of the galaxies in what is called the Doppler-Fizeau effect (the components of certain binary stars present surprising redshifts). 3) There is no uniform method for measuring the distances of different galaxies. For nearby galaxies, astronomers rely on the relationship between the absolute magnitude and the period of variable stars (Cepheids). By comparing its absolute magnitude and its apparent magnitude, one can calculate the distance of a star, which provides the distance of the galaxy to which it belongs. However, this only holds for Cepheids in nearby galaxies. For remote galaxies, the astronomer is forced to choose the most luminous bodies as indicators of distance (globular clusters, supernovae), but each of these objects presupposes a different calibration, which presents a potential source of error. 4) A recently introduced method based on radio astronomy uses the Tully-Fisher relationship between the velocity width and the intrinsic luminosity of a galaxy to deduce its distance, by comparing its absolute magnitude and its apparent magnitude. This method is not applicable, however, to spiral galaxies that contain little neutral hydrogen. 5) The discordant results of the two research groups can be attributed to their use of different methodologies. Sandage chose a small number of indicators, whereas Vaucouleurs adopted a large number of indicators and relied on calibrations. A reliable determination of the Hubble constant will only be possible with the stellar parallax method, which remains out of reach

despite the Russian Radioastron Project (Hetherington 1993; Leverington 1995; Novikov and Sharov 1995).

9. The publication by the architect-mathematician Girard Desargues of *Exemple de l'une des manières universelles touchant la pratique de la perspective* (1636) and three years later of the *Brouillon project* (1639) gave rise to a polemic in which the author and his ally Abraham Bosse of the Académie Royale de Peinture et de Sculpture found a group of powerful opponents arrayed against them. The Reverend Father Jean Dubreuil published under the pseudonym "Jésuite de Paris" a work entitled *Perspective pratique* (1642), which contained an expurgated version of Desargues' *Manière*. Desargues replied with two leaflets outlining the egregious errors contained in his critic's treatise and supported his case with accusations of plagiarism in *Diverses méthodes universelles*. Dubreuil retaliated with satirical sallies and ironical disquisitions in *Avis charitables sur les diverses oeuvres et feuilles volantes du Sieur Girard Desargues*. The architect-mathematician reciprocated with a treatise entitled *Six erreurs aux pages 87, 118, 124, 128, 132 et 134* (1642). The following year Abraham Bosse took up his pen in Desargues' defense with *Manière universelle de M. Desargues pour pratiquer la perspective* (1643), which provoked a response from J. Curabelle in the form of *Examen des oeuvres du Sieur Desargues* (1644). Desargues reacted angrily to these new attacks (Poudra, 1864) and it seems that in the end professional interests became the over-riding factor in this dispute over a purely theoretical issue.

10. Desargues' theorem states that if three straight lines joining the corresponding vertices of two triangle all meet in a point (the "perspector"), then the three intersections of pairs of corresponding sides lie on a straight line (the "perspectrix"), and vice versa. His theorem made it possible to reconstruct many of the geometrical properties of perspective and, most notably, to define in mathematical terms the *tiers poincts* or "distance points" which had previously been employed only in an empirical manner.

11. In this way, taking into account the logical structure of the problem posed by the direction of visual rays, students of optics during the thirteenth century could adopt either the intromissionist or extramissionist interpretation of vision, or a combination of the two (see Chapter 4). Another well-known example is the description of the nuclear potential; Schrödinger's model can be regarded as a "family of models" rather than a monolithic notion (Giere 1988, 184).

12. Venturing beyond the field of science in *stricto sensu*, the case of the *querelles amicales* between Jansenists is known. Once again, far from forming a monolithic block, the members of the Jansenist movement espoused such a wide range of views that they regularly held verbal jousting matches amongst themselves. These minor differences of opinion were so frequent that Antoine Arnauld proposed that they be classified as another form of controversy (Solère 1995).

13. Within the realm of scientific activity, this negative correlation is perhaps not as generalized as Coser believes. As we know, researchers seek to publish their results and to uphold the meritocratic system. Nevertheless a topical controversy offers the opportunity to win fame by resolving a problem that

has perplexed the scientific community. Given this, it is difficult to see how a size effect could condition the intensity of the members' involvement. Anyone with a relevant hypothesis would have an interest in participating in the debate.

14. And so it was that Alfred Wegener's ideas regarding continental drift, although published in 1915, were not seriously discussed until the 1950s. During the 1920s the conditions necessary for his radical thesis to give rise to controversy had not yet converged. As Ronald N. Giere remarked in 1988, the focus of geologists in that moment was light years away from the theories of Wegener. The commentary of a participant at the 1922 meeting of the Geological Society of America is illuminating in this regard. Grier says: "If we are to believe Wegener's hypothesis we must forget everything which has been learned from seventy years and start all over again" (238–239). The ideas of the German geologist were therefore not taken seriously in that epoch.

15. It may be added that the spontaneous mediators in a conflict are generally recruited from these social circles, in which case the problem can be resolved with little clamor. The unusual configuration of the dispute between Hubble and van Maanen in the 1930s regarding the rotational velocity of galaxies can be explained by the fact that they both worked at the same institute, Mount Wilson Observatory. As Hubble later recounted, when he suggested that the results of van Maanen fell "within the uncertainty of the determination" he found his criticism rebuffed "in a wildly personal manner." At the same time, an ally of van Maanen remarked that Hubble "showed a distinctly ungenerous and almost vindictive spirit." Nevertheless, this dispute cannot be considered a public controversy, because it was constantly stifled by the director of the observatory, Walter Adams, and the editor of the institute's publications, Frederick H. Seares. Seares writes in this regard: "For two men in the same institution there is opportunity for personal contact and for direct examination of each other's results, and hence for private adjustment of differences in opinion. The institution itself, it seems to me, is under obligation to see that all adjustment possible be made in advance of publication. If agreement cannot be attained it may be necessary for the institution to specify how the results shall be presented to the public" (letter dated 1935 and cited by Sharov and Novikov 1995, 85). Adams and Seares did everything in their power to convince the two scientists to resolve their disagreement privately. Hubble finally decided to ignore the contradiction between his results and those of his rival, and to keep silent.

16. Beginning in 1997, Henrik Svensmark of the Danish Space Research Institute has been publishing results which purport that global warming is due to variations in solar activity. The broad outlines of his theory are as follows. The magnetic activity of the sun, which fluctuates in cycles of approximately eleven years, gives rise to a solar wind (a continuous stream of charged particles composed principally of protons and electrons emitted by the sun) which fills the space of the heliosphere, hampering the penetration of galactic cosmic radiation [GCR] into the earth's atmosphere. The correlation observed between the flux of cosmic rays and the cloud cover (measured by the satellites Nimbus-7, ISCCP-C2, ISCCP-D2 and DMSP) provides the starting point for a causal explanation of the appearance of clouds. The last

link in this causal chain is well known: the cloud cover prevents the rays of the sun from reaching the ground, thus causing a lowering of the earth's temperature. When solar activity increases, the solar wind gains strength, the penetration of GCR diminishes, the cloud cover clears, and the climate warms (Svensmark and Friis-Christensen 1997; Svensmark 2000; Marsh and Svensmark 2001). The preprints of Svensmark's papers are available at http://www.dsri.dk/~hvs. The *Climate Gate* scandal exploded in November 2009, but with a change in the direction of the attack, which veered toward a critique of the intergovernmental organization itself. E-mails were made public that exposed the hybrid status—half scientific, half political—of the IPCC. In one of these the director of the IPCC, Philip Jones, declared his intention to bypass the rule of peer review: "Mike . . . The other paper by M.M. is just garbage—as you knew. De Freitas again . . . I can't see either of these papers being in the next IPCC Report. Kevin and I will keep them out somehow—even if we have to redefine what the peer-review literature is! Cheers." (Phil Jones to Michael E. Mann, July 8, 2004). Another researcher from the University of East Anglia, Mike Hulme, acknowledged: "It is also possible that the *institutional innovation* that has been the IPCC has run its course . . . The IPCC itself, through its . . . tendency to *politicize* climate change science, has fostered a more *authoritarian* . . . form of knowledge production—just at a time when a globalizing and wired cosmopolitan culture is demanding of science something much more open and inclusive" (*The New York Times*, November 27, 2009, italics mine).

1

Relativism and Rationalism: A Metacontroversy

It is an irony of fate that the study of controversies should itself be prey to controversy in terms of its methodology, and that it cannot develop in a coherent manner if we do not first of all clarify the terms of a long-standing meta-controversy between the relativists and the rationalists.

The *social turn* (which is, broadly speaking, a movement dedicated to the exploration of the social, practical and cultural aspects of science) has raised a sometimes quite animated debate on the interest and possibility of constructing a sociology that infers scientific knowledge from the social context out of which it sprang. Many labels have been used to designate the supporters and opponents of such analyses: modernists and post-modernists, functionalists and constructivists, rationalists and relativists.

The first coupling—modernists *versus* post-modernists—has a historical context, and implies that the debate is subject to an inter-generational conflict of interest. This was Pierre Bourdieu's (1994) interpretation of the mainspring behind the new sociology of the sciences (93), and while his analysis doubtless contains some truth, a deeper investigation of the issue shows that the questions being raised by the relativists date to long before the emergence of the *social turn*. The linkage of these two terms diverts attention from the true object of the debate.

The second pairing—functionalists *versus* constructivists—focuses the spotlight on two sociological theories that presently dominate the study of science, the latter of which assumes that everything is subject to social construction, and therefore the natural world plays no direct role in the construction of scientific knowledge. However, not only do these terms fail to capture and render immediately intelligible the grounds for the conceptual opposition, but their use in this context has been shown to be inappropriate by recent developments in the sociology of the sciences. The pairing of the two concepts is asymmetrical because the discussion almost always focuses on the question of constructivism and only rarely on functionalism.

There remains the last pairing—relativists *versus* rationalists—which in contrast to the others addresses a key question in the debate,[1] and the fact that this opposition is recognized by the leading players in the two camps—for example, Hollis and Lukes (1981), Bloor (1983), Siegel (1987), Woolgar (1988), Laudan (1990), Boudon (1994), Boghossian (2006), Schantz and Seidel (2011)—would seem to confirm its appropriateness.

1. The Guiding Notions in the Debate

The more literally minded will doubtless assume that the debate that has agitated the sociology of the sciences is of recent date, ensuing because the scientific approach that developed so rapidly since the 1970s has reached its limits and can go no further. But another possible interpretation of the phenomenon must be considered.

If one reasons on a longer timescale, the present situation echoes the discussions that set rationalists against relativists at the end of the eighteenth century and, no solution having been found then, the debate has passed from generation to generation down to the present day. This more sophisticated interpretation deserves at least to be examined. Is what we have here the same debate extending from the eighteenth to the twentieth centuries, or two different debates linked by superficial resemblances? If the latter, should one take into account the historical gap by utilizing such expressions as neo-relativism and neo-rationalism?

It must be recalled that the philosophy of the Enlightenment was not received with unanimous acclaim in eighteenth-century Europe. Other currents were being defended in the name of the *Volksgeist* by such influential thinkers as Johann Gottfried Herder, whose ideas were disseminated in France by Joseph de Maistre and Louis de Bonald.[2] Adopting a position diametrically opposed to that of D'Holbach, Helvetius and the contributors to the *Encyclopédie* of D'Alembert and Diderot (1751–72), they insisted that all knowledge should be based on common sense, local traditions, national prejudices, religious faith, and accepted authorities. As Herder (1964) writes: "Prejudice is *good*—It forces peoples to rest in their *center*, attaches them more firmly to their *stems*, to flourish *in their own way*, makes them more ardent and thus also more happy in their *inclinations* and purposes" (185). This form of protorelativism was not confined to aspects of methodology.

It led to value judgments on how civil society should be organized, and in its wake congruent political ideologies were born, ideologies that exalted the mystique of national roots and the collective imagination.[3]

It remains to be determined whether and how this form of relativism may have been transmitted down to contemporary sociologists. Aron (1967, 73, 130–131, 300), Nisbet (1984, 26), Bloor (1984, 71–94), Boudon (1986, 43) and Caillé (1989, 28) all proposed different answers to the question. There is no doubt that Joseph de Maistre and Louis de Bonald had a profound influence on the work of Henri de Saint-Simon and Auguste Comte and, through them, on the French school of sociology. All the same, this hypothesized series of links is not sufficient to demonstrate a consonance of the relativist position between the eighteenth and twentieth centuries. There was a rupture in the chain during the transmission of these influences.

Primo, the reactionary positions of the eighteenth century are not supported by today's proponents of relativism. *Secundo*, signs of a modulation in these doctrines were already detectable in the work of the first sociologists, for example, Durkheim who in the nineteenth century advanced the principle of a "secular religion."[4] If a single generation was sufficient for such inflections to appear, what form might they take during the course of six or seven generations?[5] *Tertio*, even scholars such as Caillé (1989, 28–29), who grant the influence of early relativist thought on modern sociology, recognize that the recent spread of economic models can be explained by the diminishing influence of doctrines that had previously served as a "barrier" to the rational, objective analysis of society. Relativism has therefore been propagated in inverse rather than direct proportion to the proliferation of sociological studies.

These arguments are sufficient to disprove the notion of a direct line of descent between the debates of the eighteenth century and those of today. If no nexus of transmission existed between Herder and the modern relativists, it is difficult to argue that the terms of the debate have remained the same. There exist in fact two conventional responses to this conundrum:

1. The nature of the debate between relativism and rationalism has changed since the Enlightenment, and any similarities are limited to the outward-most aspects of these concepts, or
2. The relativist notions between then and now are the same in that they constitute a reaction to rational thought, and a congruence in their conclusions can be deduced from the identical nature of their causes.

27

Of these two solutions the second appears to be more convincing, because the similarities between the relativism of the eighteenth century and the relativism of today do not end with their more superficial aspects (see Holton 1998, 49–61). It would be prudent therefore to mark this distinction, however laborious from a semantic point of view, by employing the terms *neo-relativism* and *neo-rationalism* when describing the current debate in order to avoid the risk of confusion. This would require, at the very least, an explanation of what is meant when the labels *relativism* and *rationalism* are used in this context.

1. The Relativist Approach

In the discussion that follows, the term *relativism* will be used to designate all approaches to science which envisage the possibility that social factors may have a determining influence on scientific knowledge. This semantic restriction could appear excessive, but it implies the existence of certain derived propositions that cover all the major aspects of relativism. We can see first of all that within such a framework *two distinct domains of knowledge will produce knowledge that is not only different, but also incommensurable,*[6] because knowledge is local and conventional rather than universal.

Martin Hollis (1992, 77) distinguished between "strong" and "weak" versions of relativism. In the strong version, the explication of the contents should not introduce any judgment regarding its veracity, because the notion of truth is *irrelevant* to the generation of the explanation. However, as David Bloor (1991) writes, "This poses a problem about the notion of truth, for why not abandon it altogether?" (40). From this perspective, the sociologist must consider that "instead of defining it as true belief, knowledge for the sociologist is whatever men take to be knowledge" (2). In the weak version of relativism, the notion of truth may have some meaning (i.e., as a truth that, while universalizable, is produced locally), but the sociological explanation must avoid making any pronouncements on this. It should study the content and leave the question of truth to the epistemologist. However, difficulties remain even with these stipulations.

For the purposes of this discussion it will be appropriate to adopt a more restricted definition of the term "relativism." I will not concern myself here with aspects of relativism that may manifest themselves beyond the sphere of the sciences, even though many can be found. Any conclusions with regard to cognitive relativism will therefore not be extrapolated to the realms of ethical or aesthetic relativism, despite the existence of a certain continuity between these domains. *Last, but not least,* the way the word "relativism" is used here is not intended in any way to underestimate the nuances in position that can and do exist among the relativists themselves. The debate is being conducted not only between relativists and rationalists, but also between those advocating different forms of relativism.

2. The Rationalist Approach

Just as a narrow definition of relativism can embrace various positions, "rationalism" may be described—in an equally restrictive sense—as an approach that refuses to consider scientific theories as the emanations of a constantly variable and fluctuating state of society. On the contrary, the specific mode in which scientific knowledge is constructed allows one to conclude that different societies are capable of producing knowledge that is *commensurable* because it stems from the universal ability of human beings to understand the objective structures of the world. As with relativism, however, the term should not serve as a pretext to impose an artificial unity on all forms of rationalism. The collection of studies edited by Boudon and Clavelin (1994) testifies to the existence of marked nuances in the understanding of the concept by various scholars. However, every one of their authors questions the thesis that truth can be socially determined. Moreover, one must guard against assuming that all forms of rationalism are, in the final analysis, the same. What will be explored here is rather a "sophisticated rationalism," to paraphrase Lakatos, or a "neo-rationalism" that questions how knowledge is constructed and indeed the notion of rationality itself. Boudon's research on cognitive rationality (1990, 1995) illustrates the distinction that must be made between "classical" rationalism and present conceptions of the same notion.

2. The Debate within the Sociology of the Sciences

With these semantic provisos, we will now attempt to identify the sources of the disaccord between relativists and rationalists that lies at the heart of the sociology of the sciences, beginning with a discussion of the positions held by a branch of the SSK that today has largely been taken over by the relativists.

1. The Relativist Arguments

The program espoused by the new sociology of the sciences was born of an intersection between schools of thought that embrace more or less explicitly the constructivist and relativist points of view (Lynch 1992, 215–216). It is difficult to capture the essence of the various relativist positions in a few lines, because nowhere in the literature is a succinct exposition to be found. The various positions do not even coincide with specific areas of study. The relativist attitude is highly visible in—but in no way limited to—SSK. For example, the studies of laboratory practices produced by the new field of ethnomethodology bear clear indications of an affiliation to the principles of relativism. Therefore, many variants of relativism exist and the intention here is not to list them one by one but to capture their most salient arguments.

In his review of developments in the sociology of the sciences, Michael Mulkay (1980) identifies some of the fundamental propositions of relativism. According to him, it was important to recognize that:

> Empirical observations or mathematical calculations depend for their meaning on the different background hypotheses shared by particular groups; the acceptability of a particular claim to knowledge depends on its social context; the hypotheses and the repertoires of significations used by different groups engaged in the production of knowledge are often drawn from the broader social milieu; the rules for concrete reasoning depend for their signification on informal social negotiation (126).

Mulkay notes that these conclusions, while not unanimously accepted, have been taken into consideration in many studies in the sociology of science.[7]

A somewhat extreme version of relativism is often posited in SSK studies. The most characteristic trait of this orientation is to postulate that scientific knowledge, far from imposing itself on the evidence through some mysterious adherence to reality, is actually bound by convention and is contingent in nature. This version of relativism denatures "scientific knowledge" into "scientific content" and "scientific content" into "scientific belief" in a process that effectively obliterates all distinctions between scientific knowledge and magical-religious beliefs. Woolgar (1988) states explicitly that "there is no essential difference between science and other forms of knowledge production; that there is nothing intrinsically special about the 'scientific method'" (12). This idea is deduced from a strong version of relativism, which Woolgar moreover judges to be insufficiently radical: "Relativism has not yet been pushed far enough. Proponents of relativism (both within and beyond SSK) are still wedded to an objectivist ontology, albeit one slightly displaced" (98). It is the very notion of scientific objectivity that should be banished.

The basic propositions laid out above can be arranged in a sequence that illustrates how the idea of a sociology of scientific knowledge was born:

R1 The objects of the natural world that scientific statements are related to are nothing other than "literary inscriptions" (Woolgar, Latour).
R2 The natural world plays a negligible role in the construction of scientific statements (Collins).
R3 The social context, local as well as global, plays a decisive role in the construction of scientific statements (Mulkay).
R4 Scientific knowledge is "conventional" (Bloor) and its reasoning is built on some "informal social negotiation" (Mulkay).

The Strong Program. This approach was elaborated by the Edinburgh school, represented in particular by Barry Barnes, David Bloor, David Edge and Steven Shapin. With his repeated employment of the term "scientific belief"—a use to which we will return later—Bloor unequivocally affirms his relativist position, that is that all knowledge is conjectural and relative, and reflects the situation of the person(s) who produced it. The Strong Program proposed a set of four guiding principles for SSK:

1. It would be causal, that is, concerned with the conditions which bring about belief or states of knowledge. Naturally there will be other types of causes apart from social ones.
2. It would be impartial with respect to truth and falsity, rationality or irrationality, success or failure. Both sides of these dichotomies will require explanation.
3. It would be symmetrical in its style of explanation. The same types of causes would explain, say, true and false beliefs.
4. It would be reflexive. In principle its patterns of explanation would have to be applicable to sociology itself. Like the requirement of symmetry this is a response to the need to seek for general explanations (1991, 7).

The essential point of this program is that it assigns to sociology the task of explaining scientific beliefs in a *causal* manner. Bloor (1983) furthermore states that this explanation should rely exclusively on the model of efficient causality in ascribing the causes of scientific knowledge to preexisting social conditions. As a consequence, teleological causality is banished: "How does this [teleological] model of knowledge relate to the tenets of the strong program? Clearly it violates them in a number of serious ways" (12).

The Empirical Program of Relativism. Some years later, after assessing the reactions to the Strong Program, Harry Collins of the University of Bath proposed a globally more nuanced version that took into account recent work in ethnomethodology. Nevertheless, his program remained closely tied to the principles of relativism due to the important role that he ascribed to the procedures of social negotiation. "The Empirical Program of Relativism" (Collins 1981) sets out three principles:

1. One must demonstrate the interpretative flexibility of experimental data which prevents experimentation, by itself, from being decisive. The agreement on a theory would be socially negotiated.
2. One must undertake to describe the mechanisms that limit interpretative flexibility and thus allow controversies to come to an end.
3. Finally, it would be useful to describe the detailed relationship between the constraining mechanisms of interpretative flexibility and the wider social and political structure.

Apart from a few minor differences—it may be noted that Collins does not impose a strict causal determinism on the scientific content itself—the two programs form part of a single canvas, because each begins with the presumption that truth is always local in nature, and that all truths are "ethno-truths." This fundamental position leads to a crucial question in the sociology of the sciences: how can one explain the stabilization of interpretations of the truth?

The Actor Network School. The programs of Bloor and Collins were perpetuated, and even radicalized to a certain degree, by the Actor Network Theory which was developed by Callon, Law and Rip (1986), Latour (1984, 1989), and Callon (1989), to cite just a few of the original authors. This program originated at the Centre de Sociologie de l'Innovation of the École des Mines in Paris. The fundamental notion shared by all of these authors is that the classical divisions so often used when we think about science—science and society, nature and culture, reason and belief, men and machines, etc.—should be avoided. In fact, scientific activity ought to be defined not in terms of such abstract categories, but in terms of its association with concrete elements of the widest possible diversity. Therefore, Pasteur is not a "learned man"; he is only the "spokesman" for a complex network that includes not only researchers and scientific hypotheses, but also machines and microbes, laboratory instruments and laboratory notebooks, secretaries and students, the credibility of various institutions, audiences attending public lectures, etc. The biochemistry laboratory run by Rose was picturesquely described (Law 1989) as an assemblage of rats, syringes, ether, chronometers, Geiger counters, heparin tubes, drafts of papers, proofs to correct, budget requests, etc.

Such lists show that the first characteristic of an actor-network is its *hybrid* nature. It may be observed that without such a network both Pasteur and Rose would have found themselves deprived of a field of action. The second characteristic of the actor-network therefore is its *functionality*. Finally, if one retraces the history of a scientific fact over a sufficiently long period, one will note that the network involved in its construction is not an ensemble of nodes and links that are constituted once and for all. The configuration of a participant network is *evolutive* in nature.

These observations lead to the adoption of the *principle of generalized symmetry*, because everything is subject to construction; one must be able to explain the true as well as the false, the human and the nonhuman, science and society. An approach based on actor-network theory would nullify the classical distinction between "theory" and "experience," and even more so the distinction between the "context of discovery" and the "context of justification" (Law 1989, 146–147). The representatives of the *Actor Network School* maintain that scientific facts are subject to a continual process of construction and that their "robustness" depends solely on the sociotechnical networks that have been mobilized during the course of their fabrication (Callon 1989, 30). Words such as "nature" or "society" explain nothing, because they themselves must be explained in terms of actor-networks.[8] With this realization comes a complete reversal of perspective. As Callon (1989) writes: "To accomplish this turnaround in perspective, it is sufficient to note that all the material objects on which the laboratory acts [...] constitute [...] spokespersons or representatives that, once mobilized, carry along with them all the human and nonhuman actors that they represent" (15–16).

The analysis of networks is therefore a vector of relativism insofar as it attempts to deconstruct nouns that do not have a concrete referent, including the principle of "rationality." Latour (1989) is quite explicit on this point: "Irrationality is an accusation that is always levied by the one who is constructing a network against anybody who seeks to block his way; in this way there is no coming together of minds, only networks, some longer and some shorter than others" (428). In other words, the illusion of rationality projected by certain scientific facts arises from the fact that they are associated with networks extending over vast distances that can mobilize a large number of allies to fight against concurrent interpretations.

Whether they have been linked—explicitly or not—with one or another of these three research programs, the ideas of relativism have been applied to various studies in the field of SSK. These have yielded an impressive crop of results that serve to counterbalance, if not completely deconstruct, the idealized tableau of scientific knowledge as an absolute that is always and everywhere the same, a vector of the noblest aspirations in the search for knowledge.

2. The Rationalist Arguments

The critiques that have been raised against relativism are so numerous that it would be impossible to compile an exhaustive inventory[9] and just some of the more important and recurrent objections will be highlighted here.

First critique. This addresses the supposedly innovative nature of the discoveries made by the new sociology of the sciences. Thomas F. Gieryn (1982) analyzes the constructivist and relativist program in terms of its reiteration or regression in relation to the classic sociology of the sciences. That social and cultural factors are components of all scientific activity, that all scientific knowledge is approximate knowledge, and finally that the beliefs and suppositions of scientists orient their research, are all notions that can be attributed to Robert K. Merton, although he is rarely cited by the authors who are promulgating these ideas. On the other hand, the relativist approach represents a step backward with respect to the Mertonian sociology of science. First of all, they endorse a strong version of relativism, to the point of denying that the external world contributes in any way to the production of knowledge.[10] As Collins (1981) writes: "The natural world has a small or nonexistent role in the construction of scientific knowledge" (3). Second, they confine themselves to the study of *local accounting procedures*. Karin Knorr-Cetina's research (1977, 1980) constitutes a typical example of the analysis of scientific activity based on its local social context (for example, the laboratory as a production site).

Francois-André Isambert further elaborates on Gieryn's critique. By pointing out various contradictions in the relativist program, he (1994) demonstrates that the united front presented by the SSK was based on the formation of an "anti-Mertonian coalition" rather than on any shared theoretical approach (57–65). Despite their claims, the new sociologists would have some difficulty in distinguishing themselves from their predecessors.[11] For example, Isambert (1994) shows that the first formulation of Bloor's "principle of symmetry" can in fact be found in a text by Merton (1947, 381). In a study of the reasons why the new sociology of the sciences won such rapid acceptance, after commenting on the work of Joseph Ben-David, Lewis S. Feuer and Victor C. Ferkiss, Raymond Boudon (1994) concludes that awareness of the influence of social factors on the development of science is anything but new (20). Other sociologists have voiced similar doubts regarding the relativist program, most notably during the recent reassessment of the contributions of Merton to the sociology of the sciences.

Second critique. Another objection is directed at the notion that scientific knowledge, being relative to the social situation that produced it, is no different from simple belief. Here rationalists focus on the confusion introduced by the term "scientific beliefs." David Bloor (1991) appears to distance himself from this criticism by drawing a distinction between belief and knowledge: "Of course knowledge must be distinguished from mere belief. This can be done by reserving the word "knowledge" for what is collectively endorsed, leaving the individual and idiosyncratic to count as mere belief" (5). But the criterion of collective endorsement—in itself a highly disputable notion—is purely formal, and the word "knowledge" does not appear again in his text. It is systematically replaced by "belief," a term that is problematic because it obscures an essential difference in how scientific statements are constructed in comparison to statements expressed in everyday language, a difference that Canguilhem notes (1977): "Science is a discourse subject to the norm of critical rectification" (21). The procedures for the building of knowledge are completely different from the procedures that produce magical-religious beliefs or popular opinion. No experimental protocol can be devised to prove that "Emerald is the color of the year" even if this declaration may reflect an actual consensus of opinion. Therefore, one could argue that the "consensus" which appears to give a theory legitimacy in the eyes of relativists allows them to sidestep the specificity of the clearly established procedures for the construction of scientific statements, procedures that originate in the normative dimension of scientific practice.

Another difficulty posed by the term "scientific belief" is that it leads to an "all or nothing" approach to reasoning when it comes to scientific knowledge. It is true that during the course of a scientific debate researchers will produce a series of beliefs b1, b2, b3 . . . that may even appear to be partially contradictory. But—apart from the use of the homogenous term "belief"—this is not to say that all of these beliefs can be placed on a comparable level. In the eyes of the researcher, belief b1 may represent an absurd conjecture, belief b2 a plausible hypothesis, and belief b3 a fact that has been confirmed by experimental tests. In denying such differences, one can easily arrive at the conclusion that scientific beliefs are fragile artifacts subject to endless revision, whereas in reality before they are transformed into scientific fact they must undergo a process of critical rectification and a reduction in the degree of incertitude.

Third critique. Another objection is targeted at the notion that scientific knowledge is mere convention, a point on which the advocates of relativism and constructivism have often placed disproportionate emphasis. Where does this notion come from? At the beginning of the twentieth century, Henri Poincaré and Pierre Duhem developed an epistemology that they termed "conventionalism." The current school of SSK appears to have forged a broader and even more radical acceptation of the word "convention," intended above all to demonstrate the social nature of the scientific enterprise. The differing meanings assigned to the word only partially overlap, however, and certain distinctions have to be made. Is it really the same thing to agree on *the correctness of a theory, the methodological procedure to follow,* or *the existence of a fact*? Only agreed conventions with regard to *procedures* have any chance of shedding light on the functioning of science.

David Lewis (1969) proposes strict criteria to differentiate between a contract, a convention, a norm, and a rule. A convention does not possess any inherent moral or normative qualities. It represents the outcome of a coordination of activities, leading to a tacit agreement among agents regarding the way to conduct these activities. It is "a regularity to which one thinks one ought to conform" (97). The advocates of an economy of conventions have redefined the concept of "convention" in the following manner. A convention is a "regularity in behaviour R within a population P such that:

1. All the members of the population will conform to R;[12]
2. Each member believes that all the other members of P will conform to R;
3. Each member finds in this belief a good reason to conform to R;
4. Another regularity in behavior (R') that meets these same conditions could have equally well prevailed" (Orléan 2004, 12).

Adherents to the economics of conventions often cite Hume's metaphor of the two rowers who without having to utter a word fall by tacit agreement into a shared rhythm. Each one conforms, certain in the belief that the other will do the same. This is the clearest possible illustration of a convention and there can be no doubt that similar ones exist in modern science. Today's discoveries in high energy physics are supported by the work of vast numbers of collaborators. A new record was set in 2012 with the publication of an article signed by almost three thousand coauthors (ATLAS 2012). Manifestly, countless procedural conventions had to be mobilized in order to formalize this contributorship.[13] Nevertheless, there is no intrinsic relationship between the existence of these conventions and the tenets of relativism or constructivism. It is difficult to draw from the fact of the coordinated activity of the rowers the conclusion that the river does not exist, or that the rowers have been motivated by political interest to engage in their activity. Likewise, the convention-governed nature of scientific procedures bears no relationship to the supposedly conventional nature of scientific knowledge. It is difficult to deduce from the coordinated efforts of a network of researchers that the world does not exist, or that researchers are motivated by political interests to produce knowledge.

Various authors from Duhem to Lewis have contributed to the spread of the notion of convention in the sociology of the sciences. Declarations made by Bloor (1991) testify to this:

1. That knowledge should be deemed to be crucially linked to this facet of our experience is, however, a social norm. It is a conventional and variable stress.
2. Not anything can be made a convention. The constraints on what may become a convention, or a norm, or an institution, are social credibility and practical utility.
3. Scientific theories, methods and acceptable results are social conventions (31, 43).

With regard to his first statement, one may point out that the link between a perception and a fact is indubitably the one that is least based on convention, since it is defined by an explicit protocol whose function is to guarantee the public and objective reproducibility of the experiment. The second argument fails to stand because, in comparison to experimental protocols which allow one to decide whether or not a theory is consistent with a set of data, it is difficult to see how a convention based on "social credibility" and "practical utility" can be anything but arbitrary. Against Bloor's third point, one may object that his use of the term "conventions" includes very different things, because a *theory* is not a *method*, and a method is not a *result*. Furthermore, Bloor does not back up his affirmations of any of these points, limiting himself to citing the example of the *theory* of phlogiston. Most relativists, in justifying the role of social conventions, point by turns to their coercive nature and to the procedures by which knowledge is constructed.

Primo, certain sociologists defend the thesis of the conventional nature of scientific knowledge by observing that all human practices rest on "socially instituted rules" (Fourez 1996). They then attempt to demonstrate the social character of objectivity by drawing an analogy between scientific research and the schoolroom exercise of the dictation. When the teacher corrects a pupil's errors, "What gives this correction its quality of objectivity is its conformity with a socially accepted rule" (34). The willingness of the student to accept the corrections of the teacher is rooted in the fact that both admit the existence and authority of conventional rules, in this case of grammar and orthography. But are the practices in the schoolroom and the laboratory really comparable? The analogy is questionable because the rules of orthography in French are often arbitrary (for example, the Latin prototypes of château [castle] and honnête [honest] are *castellum* and *honnestus*, but the prototype of infâme [infamous] is *infamis*) and, contrary to the formal languages of mathematics and the sciences, natural language is polysemic and requires the manipulation of words with multiple meanings. The teacher, when correcting the pupil who has written *infame* instead of *infâme*, can only justify this act by citing convention or referring the pupil to a dictionary. In contrast the professor of physics, when correcting the equation of the center of mass $\Sigma m_i GM_i = 1$ to $\Sigma m_i GM_i = 0$, can advise the student to consult a textbook or even to conduct an experiment with weights, applying the fundamental laws of physics to demonstrate that the first equation is erroneous. Physical descriptions cannot be reduced to an interplay of social conventions.

Secundo, in support of their argument that scientific knowledge is conventional in nature, relativists claim that scientists are actually engaging in the "social negotiation of knowledge" when they conduct informal discussions. In their view, such negotiations are to be taken at face value as they are capable of determining the truth of scientific pronouncements: "*Negotiating* knowledge almost always signifies modifying its content [. . .]. [Scientists] fabricate in this way *compromises* that affect the contents themselves of this knowledge" (Callon and Latour 1991, 30, italics mine).

Rationalists would argue that it is—to say the least—dangerous to speak of a compromise between two scientific theories in the same way that one might refer to a compromise between the opposing proposals of trade unionists and their employers (justifiable in the latter case as it involves the negotiation of a compromise between initial propositions that are not subject to the criterion of *truth*). While relegating this image of workers *versus* bosses to the category of metaphor, rationalists do acknowledge the existence of discussions that *accompany* the search for the truth. However, the free exchange that takes place is perhaps only a reflection of the scientists' hesitation before choosing, among various concurrent interpretations, the hypothesis that presents the best chances of being true.

In a critique of the relativists' classification of the various modes of resolving controversies, Ruth Macklin (1987) shows that the negotiated settlement does not exist in *stricto sensu* in the case of scientific controversies. Robert F. Rich (1987) arrives at the same conclusion at the end of a comparative analysis of political and scientific controversies: "It is worth noting that in science one is not dealing with a 'negotiated settlement.' Instead, by using well-developed canons of research and evidence, a procedure is legitimately developed for the settlement of disputes without having to rely on negotiation. Scientists characterize this process as a 'rational' one" (162). For his part Raymond Boudon (1990) chose to analyze controversies that, given their long duration, might create the impression that some form of negotiation had taken place:

> Like the debate between Millikan and Ehrenhaft, the debate over the theory of evolution has been eternalized. [. . .] But the twists and turns that characterize a scientific debate such as this one are no different in nature from those that accompany a judicial investigation. *For as long as the investigation lasts*, the partisans of the two concurrent theories T and T? [. . .] often have good reasons—that is to say, reasons which may not be objective, but no more are they necessarily arbitrary—to adhere to one or the other (224–225).

This reading contrasts markedly with the ideas of Bloor (1983), Collins (1981), Woolgar (1988), and others, who maintain that the truth of a theory is by its very nature relative because it is based on negotiations within the scientific community. From the rationalist's perspective, the truth of a theory is not determined, but may be *accompanied* by a social negotiation that nevertheless remains strictly incidental to the process. If this is the case, does sociology have anything relevant to say with regard to scientific knowledge? Some, such as Isambert (1994), do not deny that scientific knowledge could constitute a subject for sociological study, but it would be important to do so without adhering to an overly narrow relativist position with its sometimes excessive emphasis on the conventional nature of scientific knowledge.

Fourth critique. Rationalists also object to the widely-held idea among sociologists that there are no criteria to demarcate science from nonscience (and, notably, science from society). Latour (1989), for example, refuses to separate the inside and the outside of science based on his critique that rationalism and sociologism are mistaken to place scientific and non-scientific factors in separate compartments (434).

Benjamin Matalon (1986) notes that the principle of nondemarcation perpetuates the use of the expression "scientific belief" leading to an obscuring of the specificity of different human activities, each of which has its own stock of rules and behavioral norms. As others have observed, "The partisans of the 'Strong Program' declare that society and science form an inseparable whole, but in practice one can clearly distinguish between developments that are 'internal' and those that are 'external' to science" (Lécuyer et al. 1988, 406). Lécuyer's critique, however, was aimed specifically at the papers published in *Les scientifiques et leurs alliés* (1985).

All the same, in his examination of the controversy between Pasteur and Pouchet, Latour reintroduces the demarcation between science and nonscience, in this way rendering ambiguous the notion that science may be subject to all manner of "external influences," because he is here reviving the dichotomy between "interior" and "exterior" that he previously claimed did not exist. On a factual level, one can observe this ambiguity in the supposed influence of Pasteur's political choices on the outcome of his dispute with Pouchet. Latour (1989) writes: "Here we see an excellent ideology, of gigantic proportions" (437). Yet this conclusion, which places theory in the position of *explanandum* and ideology in the position of *explanans*, makes sense only if a demarcation is in fact endorsed between science and ideology. The classic distinction between epistemic and nonepistemic factors, or any other division of this type—such as the distinction between "standard factors" and "nonstandard factors" proposed by McMullin (1987, 61)—thus remains pertinent to the analysis of how science is produced.

Fifth critique. One of the most important objections to the relativist position relates to its insistence on the causal determination of knowledge. The principal of causality in the Strong Program requires that one "investigate and explain the very content and nature of scientific knowledge" exclusively from the perspective of "efficient causality," in that it stands in opposition to the teleological model (Bloor 1991, 3, 12).

This position must be understood in *stricto sensu*; it is the mechanism of social determination that induces Bloor (1991) to subscribe to the theory of social-historical determinism: "To believe that ideas are determined by social milieu is but one form of believing that they are, in some sense, relative to the actor's historical position" (15).

On this question Bloor places himself squarely in the tradition of Durkheim regarding the social origin of concepts. Applying these principles to his study of the science of mathematics by the ancient Greeks, Bloor shows that their conception of numbers differed profoundly from our own. Incommensurable magnitudes such as $\sqrt{2}$ were not viewed as numbers. However he ascribes this difference to the specific state of society in ancient Greece, with the understanding that the "Grecity" of this mathematical construct should be considered as an irreducible singularity. Raymond Boudon (1994) has demonstrated that a rationalist would have drawn a completely different conclusion from the same set of facts: "It is certainly not the state of Greek society but the state of mathematics at the time that accounts for this interpretation. And one can wager [. . .] that it would have been identical in Monomotapa, if their mathematics had been the same" (31).

The presumed influence of culture on the conception of mathematics is subject to yet another criticism: Did the Greeks really think about numbers in a uniform manner? True, a debate was being conducted on the nature of the diagonal of the square, but this discussion was limited to the school of Pythagoras. In contrast, other Greek mathematicians sought to bridge the gap between rational numbers and irrational numbers.

In his *Measurement of a Circle*, prop. 3, Archimedes provided the upper and lower bounds of the value of π in terms of the circumferences of similar inscribed and circumscribed polygons: the greater the number of sides to a polygon, the more accurate the estimate of the numerical value of π based on this polygon. Archimedes stopped at ninety-six sides (three decimal places of accuracy) but he could have gone much further. Later, using the same method al-Kāshī progressed from three to seventeen decimal places of accuracy.

In his *Arithmetica*, Diophantus laid the cornerstone for the subsequent writing of a number α_0 as a continued fraction. The number α_0 is *rational* if it can be written as a *simple* continued fraction (that is, if the division stops), whereas it is *irrational* if it is written in the form of an *infinite* continued fraction (that is, if the division never stops).

41

In point of fact, Archimedes and Diophantus were themselves Greek. Therefore, the thesis that "the Greeks" as a whole were resistant to the notion of irrational numbers and only accepted the existence of integers and rational numbers due to their "Grecity" holds true only if one limits Greek mathematics to Pythagoras, which would clearly be an absurdity.

Bloor's methodological relativism owes much to Émile Durkheim, most notably the French sociologist's idea of the "social determination of knowledge."[14] In *The Elementary Forms of the Religious Life*, Durkheim (1912) sustains that "The concept of totality is merely the abstract form of the concept of society: it is the whole that includes all things, the supreme classification that encloses all other classifications" (630). As a consequence, all of our concepts are the result of our capacity to formulate abstract totalities—of space, time, numbers, and so on. The question that concerns Durkheim is to identify where this faculty comes from. His response is that man's ability to forge and use concepts stems from his earliest experiences of collectivity, which would have furnished him with an example of the subsuming of the particular (the individual) to the general (society), a prototype from which he might then deduce all other concepts.

However, the rationalist could equally well reverse Durkheim's sociological interpretation and suggest that man is innately endowed with certain cognitive abilities that allow him to formulate abstract totalities, which he then applies—sometimes to the social experience, in this way establishing the concept of "society," and sometimes to the physical experience, determining in an independent manner the concepts of "force," or "space," or "time." This would constitute an equally satisfactory causal analysis, but one that sweeps away the thesis of the social origin of concepts. Another objection to the "holistic" sociology of Durkheim, raised in this instance by Piaget (1965), argues that the link between "force" and "*mana*" is not one of continuity as postulated by the French school, but a process of "decentration" of the concept vis-à-vis belief (70). Sociologists who ignore this relationship of decentration will end up confounding ideologies and symbolic beliefs, which are imposed by social constraint, and scientific knowledge, which is the outcome of a mechanism of "operating coordination."

What is more, a causal analysis would require that we set aside all uncertainty regarding the direction in which the causal arrow is pointed. Clearly, from a nominalist perspective it would be more prudent to define "Grecity" in terms of the ensemble of works produced in ancient Greece (of which a number are mathematical in nature) rather than trying to explain Greek mathematics in terms of its Grecity. The fact that some scholars explain mathematical works by their Grecity whereas others reduce Grecity to the works produced at this particular time and place shows that the antecedent and consequent are interchangeable. The causal model therefore rests on dubious foundations. As will be shown in the following chapters, it is often difficult to prove that religious or ideological connotations preceded rather than followed a scientific theory, thus opening the way to the placing of excessive weight on social factors in the determination of knowledge.

Sixth critique. Another crucial objection can be raised against the notion of a causal determination of scientific knowledge. This was clearly enunciated by Mario Bunge (1991): "Causal analysis is not enough and moreover it is sometimes inappropriate, because it focuses on external conditions, ignoring the inquirer's motivations and cognitive problems, or attributes them all to external factors, thus making him appear as a mere pawn rather than as a creator" (536). If one takes the idea of causality seriously, the constructivist approach to the sciences raises an almost insoluble problem. And this is the reason: 1) Certain causes that are supposed to determine scientific knowledge (such as political ideologies, religious dogmas, etc.) apply to a very broad population; 2) but under the system of "efficient causality," which implies determination, the same causes must produce the same effects; 3) as a consequence the individuals making up this population must produce, with a similar impetus, knowledge that is in conformity with the dominant ideology or social structure. But such is not the case, because in fact there are enormous differences between the mental content of an architect, a musician, and a mathematician working in the same period, and equally marked differences between the mental content of two scientists working at the same time, even when they belong to the same discipline.

Sociologists who espouse the notion of the causal determination of content must explain why the same set of social causes S can, in a given society, generate different sets of cognitive content c_1, c_2, c_3 . . . .[15] Examples are certainly not lacking. Why did the social context of France in the early 1900s engender mental content as different as [c1] the study of the three-body problem by Henri Poincaré and [c2] *Alcools* by Guillaume Apollinaire? The difficulties encountered in trying to explain these differences on a macro level are reproduced on the intermediate level, for example in the field of physics. Where does the similitude lie between [c3] the discovery of gamma rays by Paul Villard, [c4] the study of chromosphere and solar prominences by Henri Deslandres, and [c5] the works on special relativity by Paul Langevin? Nor does this problem vanish on the micro level; it is well known that the Lumière brothers collaborated closely, but it was Louis who invented [c6] the trichrome process and not Auguste, who instead is credited with [c7] several discoveries in biology. Finally, the objection remains even on the individual level; what, for example, is the homology in content between [c8] the first article on special relativity, *Zur Elektrodynamik bewegter Körper*, and [c9] the light quantum hypothesis? These theories were proposed by the same author (Einstein) in the same place (Bern) in the same year (1905), but they present content that physicists today would consider to be contradictory. Is the explanation really that Einstein was subject to the influence of two different societies simultaneously?

With regard to the thesis of the determination of knowledge, there is a simple explanation for the difference in works produced at the same time in the same discipline. It is a fact that a researcher will address *specific problems*, whose formulation depends on how the objective world is structured. And yet this conclusion is dismissed by relativists,[16] who claim that the natural world plays no role in the production of scientific statements (Collins 1981; Woolgar 1988; Latour 1988). SSK therefore must either abandon the relativist thesis or demonstrate that the apparent differences between the works of Ravel and Poincaré or between Einstein-1 and Einstein-2 can be reduced to a common denominator. The hypothesis of "homologies" never having led to incontrovertible results in sociology,[17] we must accept the idea that the scientific content produced in the same period can vary because it corresponds to different structures in the objective world.

If the points raised above are arranged in order of importance, the crucial nature of the question regarding the *ahistoricity of the norms of rationality* becomes immediately clear. It is a question that pertains to the domain of epistemology, which I will not attempt to address for the moment, merely observing that, despite the support for the relativists' views in contemporary thought, epistemology has produced powerful arguments against them.[18]

The relativist position in favor of the historicity of the norms of rationality conditions at one and the same time the idea of scientific beliefs, their fragility, the role of social negotiation between researchers, and—last, but not least—their determination by extrascientific factors. For the rationalist, the acceptance of a theory depends first of all on the criterion of proof, that is, on the researcher's ability to establish the correspondence between a theory and his experimental results, and to produce a demonstration in conformity with the rules governing scientific evidence, one that will lead to a lasting consensus within the scientific community, irrespective of ideological, political or religious considerations. For the relativists, on the contrary, the norms of rationality are neither absolute nor universal. The consensus that forms around a theory is not logical in nature, but *socially constructed.* Furthermore, the relativists endeavor to cast doubt on the very concept of "truth." It is not that they interpret truth in an idiosyncratic manner; some use the concept of truth in a very classic manner (Bloor 1991, 41). It is more that in their view the norm positing that a truth must correspond to the experimental results is vague and, what is more, inconsistent. Based on a specific reading of the works of Duhem and Quine, Bloor (1991) affirms that "This is not the correspondence of the theory with reality but the correspondence of the theory with itself [. . .] The process of judging a theory is an internal one" (38–39). At this point he could very well ask himself whether the notion of truth is of any utility at all in science, and in fact, as was noted above, Bloor writes "Why not abandon it altogether?" (40). Another proponent of the (presumed) brittleness of the objective world, Steve Woogar (1988), set out an equally radical definition of truth, in which he posits that the real is opaque to any form of investigation: "Representation constitutes object. The idea of reversing the arrow is to suggest that objects are constituted in virtue of representation [. . .] Facts and objects in the world are inescapably textual constructions" (56, 69).

In the eyes of the relativist, the scientist is therefore a prisoner of his own monologues, providing the world with properties that are—if any—fictitious, and that in reality spring from the social context in which he lives. The gulf between the relativists and the rationalists is at its widest here, and we can now consider how the study of controversies might shed some light on the debate.

3. Relativists versus Rationalists: The Place of Controversies

If one compares the guiding principles of the relativist programs of Bloor (1983) and Collins (1981), it is clear that a number of the points enunciated by both are in agreement on the questions raised by the sociological study of scientific controversies. They posit that during the course of a debate one can in fact observe: 1) the transformation of certain initial hypotheses to the status of accepted explanations;

2) the corresponding rejection of alternative hypotheses, now classified as nonconforming explications; 3) a fluctuation in the positions
of the protagonists; and 4) the influence of extrascientific factors
on these positions. In addition, scientific controversies are of great
interest to sociologists because they can in these cases apply "the
regard of the outsider," as described by Shapin and Schaffer (1989).
In effect, "Historical actors frequently play a role analogous to that
of our pretend-stranger: in the course of a controversy they attempt
to deconstruct the taken-for-granted quality of their antagonists'
preferred beliefs and practices, and they do this by trying to display
the artifactual and conventional status of those beliefs and practices"
(1989, 7). The sociological interest of debates involving contradictory positions therefore lies in the fact that they reveal elements and
mechanisms which generally remain tacit when science is pursued in
a nonconflictual fashion.

But the most interesting sociological questions raised by the study
of controversies focuses on the procedures used for their settlement.
According to the schema proposed by SSK relativists, the resolution
of a dispute depends on the outcome of an uninterrupted series of
complex negotiations and the accumulation of asymmetries between
rival theories, in which it is understood that the elements mobilized by
the parties for and against a given hypothesis will consist not only of
rational arguments, but also a range of professional, economic, political,
and even axiological interests. Latour (1989) suggests that an analysis of
this process might begin with a list of the elements that upset the initial
symmetry between two rival scientists, with corroborating evidence
accumulating more rapidly on one side than the other. Such an analysis
would require that all the asymmetries be taken into account, no matter
what domain they may belong to, and therefore should include not only
factual "proof" but also personal convictions, professional reputation,
the ties of a researcher to a commission of experts, or even his possible
collusion with political or economic lobbies.

The interest of scientific controversies resides as well—in broader
programmatic terms, because no "rationalist program" in *stricto sensu*
exists in this area—in the fact that, among the observable scientific
phenomena, controversies provide the best practical laboratory for
the critical scrutiny of the tenets of relativism. In point of fact, one can
test the validity of various readings in relation to the elements of reality
selected and in relation to the sequence of propositions that unfolds
[*l'enchaînement des propositions*].

Primo, considering that a procedure must always begin—if only due to limitations of time and space—with the selection of the most salient elements of a problem, one must obviously ask oneself what the criteria for such a selection should be. It may be noted that works in the field of SSK rarely pose this question. *Secundo*, any attempt to establish the deterministic relationship between scientific knowledge and its social context requires that we verify the pertinence of causal interpretations. Here again, little effort has been made to develop such criteria in SSK studies. *Tertio*, sociologists have attempted to show that scientific inquiries conducted in another time or another place do not correspond in any way to the norms of contemporary science. But it then becomes necessary to question our capacity to understand the results of past scientific endeavors (as is attested to by the daily research of the historian of science).

Given the universality of the scientific controversy as a social phenomenon down through the centuries, it also becomes necessary to ask the more general question: on what is the persistence of this largely transhistoric and transcultural activity based? Scientific controversies are not in any way specific to modern science. Debates regarding the construction of the regular heptagon, the trisection of an angle, and the duplication of the cube arose with the ancient Greeks and were revived in medieval Islam, as is testified to by Abū al-Rayḥān al-Bīrūnī (see Chapter 5).

Having defined the framework of this inquiry and laid out the issues that a study of controversies may address, we will now undertake a series of empirical investigations into specific scientific controversies. The next chapter will examine the most celebrated controversy of all—that between Louis Pasteur and Félix-Archimède Pouchet over the thesis of spontaneous generation.

Notes

1. We will set aside the opposition between constructivism and realism. Ian Hacking (1992), who defends certain aspects of constructivism, writes in a study on statistical knowledge: "Is the result some kind of 'relativism' about truth and styles of reasoning, some kind of 'anti-realism'? No. That by which we investigate reality is not relative to anything, and the aspects that we call the real determine what is true or false according to our criteria. Yet our styles and our truths do not exist until we bring them into being" (1992a, 155).
2. The most important works produced by this school of thought are: Herder, *Ideen zur Philosophie der Geschichte der Menschheit* (1791), *Briefe zur Beförderung der Humanität* (1797); Maistre, *Considérations sur la Révolution*

française (1796), *Essai sur le principe générateur des constitutions poli-tiques*(1808); and Bonald, *Théorie du pouvoir politique et religieux* (1796), *Essai analytique sur les lois naturelles de l'ordre social* (1800).

3. In direct opposition to the relativists may be placed the universalism of a thinker such as Montesquieu (1941): "If I knew something useful to myself and detrimental to my family, I would reject it from my mind. If I knew something useful to my family but not to my homeland, I would try to forget it. If I knew something useful to my homeland and detrimental to Europe and Mankind, I would consider it a crime" (10).

4. We could retrace this inflection even further back, citing the existence of an "earliest French school of sociology," which is little known today because of the strong mathematical orientation given it by its founders. We may mention the work of authors such as Maupertuis (1698–1759), Condorcet (1743–94), Monge (1747–1818), Laplace (1749–1827) and Poisson (1781–1840), all of whom were interested in the theory of probabilities and who made an important contribution to the sociological theory of the formation of judgments, beliefs and prejudices. It goes without saying that these works do much to counter Caillé's argument cited in *tertio*.

5. Bloor (1991) perceives this problem clearly: "The question, now, is why this repeated pattern of ideological conflict [between Enlightenment and Romanticism] crops up in an esoteric area like the philosophy of science?" He responded by arguing for the influence of ideology on science: "The hypothesis that has already been advanced to predict and explain this similarity is that theories of knowledge are, in effect, reflections of social ideologies" (74–75).

6. The term "incommensurable," which was introduced by Kuhn (1962) and Feyerabend (1975), is often employed by sociologists who support relativism; see Callon and Latour (1991, 23).

7. For example: Bloor (1973, 1976), Collins (1974, 1981), Collins and Cox (1976), Norton (1978), Collins and Pinch (1979, 1994), Shapin (1979), Barnes and Bloor (1981), Callon (1981), Lynch (1985, 1993), Galison (1987), Latour and Woolgar (1988), Shapin and Schaffer (1989), Woolgar (1988), Latour (1989), Callon and Latour (1991), MacKenzie (1993), Vinck (1995), Knorr-Cetina (1995), Barnes (1997), Restivo (2005), Akrich et al. (2006), Moore and Stilgoe (2009), Venturini (2010), Besel (2011), Bechler (2012), Godwin (2013), etc.

8. We may cite here two of Latour's (1987) "rules of method": Rule 3: "Since the settlement of a controversy is the cause of Nature's representation, not its consequence, we can never use this consequence, Nature, to explain how and why a controversy has been settled." Rule 4: "Since the settlement of a controversy is the cause of Society's stability, we cannot use Society to explain how and why a controversy has been settled" (258). This approach has been implemented in various studies on scientific controversies, see for instance Callon (1981), Latour (1989), Callon and Latour (1991), Kim (1994), Venturini (2010), Besel (2011).

9. To mention just some of the most significant critiques of the relativist position: Desanti (1975), Hollis and Lukes (1981), Laudan (1981), Gieryn (1982), Freudenthal (1984), Isambert (1985), Matalon (1986), Siegel (1987), Lécuyer (1988), Boudon (1990, 1994), Laudan (1990), Radnitzky (1990), Ben-David (1991), Chateauraynaud (1991), Bunge (1991/2, 1999, 2006, 2012), McMullin

(1992), Gross and Levitt (1996), Cole (1996), Sokal et Bricmont (1997), Koertge (1997), Dubois (1998), Bouveresse (1999), Boghossian (2006), Berthelot (2008), Schantz and Seidel (2011). It is only in recent times that the rationalist approach has been tested in studying scientific controversies. Apart from my articles of 1998 and 1999, see Machamer, Pera, and Baltas (2000), Li (2011), Junges (2013).

10. This type of constructivism is in reality far removed from the constructivism of Pierre Duhem, who never disowned the correspondence and consistency theories of truth, even if a more radical emphasis can be detected in places. For example: "Suppose, for instance, that it is a question of an experiment on the phenomena of optical interference. The report of such an experiment contains statements bearing surely on the objective characteristics of light [. . .] Now, we must not attribute offhand complete and entire objective reality either to vibrations of an elastic ether or to polarization of a dielectric ether, for they are really symbolic constructions imagined by theory in order to summarize and classify the experimental laws of optics" (Duhem 1993, 443).

11. For example, in *Laboratory Life* Latour and Woolgar (1979, reed. 1986) sought to distance themselves from the classic approach by placing emphasis on the *constructed* aspect of science, even though: 1) this position has long been recognized by sociologists and science historians; 2) the authors infer from this, by a sort of paralogism, the notion that the external reality corresponding to scientific pronouncements does not exist independently from its construction: there exist only *artifacts*, and "'out-there-ness' is the consequence of scientific work rather than its cause" (1986, 182).

12. This proposition summarizes the thesis of Lewis (1969).

13. Such forms of coordinated scientific research have been studied in detail in the case of the determination of the speed of light (Raynaud 2013).

14. The use of the words "to determine" and "determination" has a long history in philosophy (Hegel, Spinoza, Marx, etc.). Durkheim (1937) employed these terms when seeking to define how social facts should be explained: "The *determining* cause of a social fact must be sought among antecedent social facts and not among the states of the individual consciousness" (109, italics mine). The notion of "determination" must always conform to the sense of a necessary causal relationship. Durkheim nevertheless cannot be considered a precursor of the school of SSK, because nowhere does he claim that collective representations of the truth are determined by the state of the society. On the contrary, collective representations "have the power to attract and repel each other and to form amongst themselves various syntheses, which are *determined* by their natural affinities and not by the condition of their milieu [. . .] They are not directly related to particular features *determined* by the social morphology" (Durkheim, 1963, 34, italics mine). In this passage Durkheim explicitly asks us to consider the internal order according to which representations are structured rather than the influence that the social milieu might exercise on them. Bloor (1997, 58) is therefore correct to insist that Durkheim's approach does not stem from Marx, although his interpretation of the French sociologist's texts remains debatable.

15. We observe that sociologists must equally be able to explain why two distinct ensembles of social causes S and S' are able to produce identical mental content c. This is the essential problem posed by the "precursors"

in the history of the sciences. How can we explain the phenomenon of the independent discovery by different scientists of the same theory? How can we explain the fact that scientists separated by vast distances in terms of both space and time may have been working on the same problem? We can cite here the long series of savants who sought to find an explanation for the phenomenon of refraction: Ptolemy, Ibn Sahl, Ibn al-Haytham, Bacon, Harriott, Snell, and Descartes.

16. It may be noted that Bloor (1991) establishes the notion of the relative nature of theoretical constructions in relation to phenomena, without negating the existence of reality (39).

17. The thesis of the art historian Erwin Panofsky (1967)—who saw structural homologies between Gothic architecture and scholasticism in the tripartitioning of the *triforium* of the Gothic cathedral and the tripartitioning of philosophy into *partes*, *membra*, and *articuli*—is based on a doubtful parallel. With regard to the presumed tripartite construction of the arched galleries of cathedrals, we may cite almost as many two-fold (Vezelay), three-fold (Amiens), four-fold (Chartres) or six-fold (Bourges) arched galleries. As for the supposed tripartite structure of medieval treatises on philosophy, this was in fact anything but the rule (consider, for example, Roger Bacon's *Opus maius*). The sociology of the arts has largely turned away from this approach (Pedler 1998, 195), although Panofsky's observations do apply *a fortiori* to scientific content, which are less equivocal than artistic content.

18. The two arguments of Siegel (1987) in this regard may be cited. The first is the UVNR argument [relativism "undermines the very notion of rightness"]: "Relativism is incoherent because, if it is right, the very notion of rightness is undermined, in which case relativism cannot be right." The second is the NSBF argument ["necessarily some beliefs are false"]: "Relativism is incoherent because it holds that all beliefs and opinions are true, yet, given conflicting beliefs, some beliefs must necessarily be false, in which case relativism cannot be true" (4–6). Siegel shows that all the arguments marshaled in defense of relativism must necessarily fall before these two arguments UVNR and NSBF. Therefore, the conception of "relative truth" is self-refuting. A general overview of the research being conducted on the notion of "truth" is provided by Engel (1989, 1998).

2

The Controversy between Pasteur and Pouchet: An Essay on the Principle of Accumulated Asymmetries

Various cases have provided the occasion for an exposition of the principles of SSK, the most celebrated of which concerns the theory of spontaneous generation. This debate was between Félix-Archimède Pouchet (1800–72) and Louis Pasteur (1822–95) in the years 1859–64. The similar conclusions arrived at by various scholars—Farley and Geison (1974), Mendelsohn (1987), Latour (1989), Lécuyer (1990), Collins and Pinch (1993), and Vinck (1995)—give the impression that the debate over spontaneous generation constitutes an exemplary case, all of whose characteristics confirm the principles of the relativist program. Nevertheless, some science historians, such as Roll-Hansen (1979), Gálvez (1988), and Crosland (1992), have arrived at very different conclusions on the same subject. Even certain sociologists question the relativist analysis of this controversy; for example, McMullin (1987, 87) and Bunge (1991, 62) express reservations as to the SSK interpretation of the political elements to the story.

Given the persistent divergence of views on this controversy, it would appear that there is still some light to be gained from a detailed examination of the debate between Pasteur and Pouchet. The case deserves renewed attention for three reasons.

Primo, this interest is motivated by the desire to follow the principles of the relativist analysis as far as possible, although "without abandoning," as Bruno Latour (1992) writes, "the solid ground of the study of empirical cases of scientific practice, the ground [that] constitutes the only stable certitude in this new field" (280).

Secundo, it is motivated by an apt critique by Terry Shinn (1994), "The quality of the studies [. . .] analyzing scientific controversies has hardly been addressed. And yet it should be noted that the descriptions to be found there are sprinkled with theoretical prejudices and, frequently, a deliberate ignoring of the nuances" (79).

Tertio, taking a fresh look at this controversy is justified by the fact that the "symmetrical" treatment proposed by certain sociologists does not reflect the actual asymmetries that existed between the protagonists and in the development of the controversy.

Thus, Farley and Geison (1991) devote twenty-eight pages of their analysis to Pasteur, and a mere ten pages to his rival in Rouen. The opponents do not receive the same amount of attention, no doubt in part because the documentation and correspondence relating to Pasteur are infinitely more accessible than the material pertaining to Pouchet. Given this disparity, there is a risk that any conclusions will consist essentially of a *critique* of the work of Pasteur. This presents a clear invitation to take up once again an analysis of the controversy and to consecrate, as Cantor has suggested (1992) much more attention to the biologist from Rouen. Given the volume of material in the archives (Appendix 2), my evaluation will be based on a selection of documents that will be cited between brackets [x] (Appendix 3). Such a historical analysis is necessary to determine the degree of validity of the relativist interpretation of the debate over spontaneous generation.

1. A Chronicle of the Controversy

Spontaneous generation may be defined as "the production of a new organized being, devoid of parents, and of which all the primordial elements have been drawn from the ambient material" (Pouchet 1859, 1). The aim of the debate was therefore to decide between the two following hypotheses: a living being is always the issue of parents (homogeny) or a living being is not necessarily the issue of parents (heterogeny). Experiments were designed to test these hypotheses, which consisted in general of studying the evolution of a sterilized fermentable milieu in a glass flask with the aid of a microscope that could detect the presence of microorganisms. The controversy rested fundamentally on the interpretation of the results. One side was firmly convinced that the microorganisms in the hermetically sealed flasks were the product of spontaneous generation, while the other side argued that they were the result of contamination.

Pouchet supported the thesis of heterogeny. In 1859 he published a book entitled *Hétérogénie, ou Traité de la génération spontanée*, in which he presented the results of experiments carried out on infusions of plant material and his explanation of these results (Plate 1). His work came to the attention of the French Academy of Sciences, but it judged the matter still open to debate and decided, in a meeting held on January 30, 1860, to offer an award—the Alhumbert Prize—of 2,500 francs to whomever could "shed new light by well conducted experiments on the question of so-called spontaneous generation." In 1859 Pasteur was not yet working on this problem, but his studies of lactic fermentation led him to hypothesize that the microorganisms in Pouchet's flasks came from the surrounding air. Pasteur had observed in his own experiments that the solution remained unaltered if it was isolated from the atmosphere or exposed only to calcined air. Pouchet read Pasteur's note on lactic ferments which appeared in the *Comptes rendus de l'Académie des Sciences* on February 14, 1859, and took the initiative of writing to the author. This marked the debut of the controversy. In his response dated February 28, Pasteur gave it as his opinion that the experiments conducted by Pouchet were not conclusive and suggested a different experimental approach that might allow his colleague to correct his conclusions [6*]:

> If, Sir, you adopt the arrangement that I indicate to you, in less than a quarter of an hour you could set in train a series of experiments, and you will then become convinced that in your recent experiments you inadvertently introduced ordinary air and the conclusions that you arrived at were not founded on irreproachably exact facts (1951, 46).

According to Pasteur, this experiment would be capable of proving that life is not self-engendered, but the issue of microorganisms from the surrounding air. Pouchet responded with the same argument that Needham had used against Spallanzani, that sealing the flasks created an environment that was too unlike the natural conditions required for the creation of life. There would be nothing to show whether the fermentation was due to germs contained in the air or to the chemical composition of the air itself. After all, he argued, the organic material in the infusion could very well have organized itself spontaneously into a living being under the influence of oxygen. It was therefore necessary that the solution be exposed to air, introduced in as pure a form as possible. Pasteur acknowledged this argument, but devised two sets of counterexperiments.

In the first, which was conducted in February 1860, he sought to distinguish between the effect of pure air and the effect of germs contained in the air using two different methods. He first of all repeated the experiments of Theodor Schwann (1837), and found that a solution exposed for an extended period of time to calcined air showed no signs of the development of microorganisms. He then conceived a method based on the use of "swan-necked flasks." After filling twenty glass flasks with an infusion of brewer's yeast, he had a glassblower join a long, curving glass tube whose opening did not measure more than a few millimeters in diameter to the necks of ten of them. In this way he could be certain that the composition of the air inside and outside the flasks would be identical, but it was unlikely that the microorganisms in the atmosphere would be able to travel along the entire length of the tube and reach the solution. After bringing the contents of all twenty flasks to a boil, he left them—unsealed and exposed to the air—to incubate. Forty-eight hours later he found that single-celled organisms had multiplied only in the ten conventional flasks, while the contents of the ten swan-necked flasks remained unaltered, their sinuous necks having protected the solution from the germs in the air.

To test his germ theory further, in a second set of experiments Pasteur exposed flasks containing the same solution of brewer's yeast to air at high altitudes. The atmosphere at such heights being practically devoid of microorganisms, exposure of the infusion to air would disprove the notion that oxygen alone was responsible for the appearance of life. The results of his experiments bore him out. Pasteur prepared sixty glass flasks containing a solution of brewer's yeast, heated them to sterilize the infusion, and then carefully sealed them. In September 1860, he opened twenty of these flasks near the town of Arbois at an altitude of 300 m (985 feet), allowed air to enter and then sealed the flasks. After a short period eight of them showed the presence of germs. He then opened twenty other flasks at an altitude of 850 m (2,800 feet) on Mount Poupet; after a short incubation period five flasks were found to be contaminated. Finally, he opened the twenty remaining flasks at an altitude of 2,000 m (6,500 feet) close to the Mer de Glace glacier on Mount Blanc and only a single flask became contaminated. In this way Pasteur demonstrated that there was an inverse relationship between altitude and the presence of microorganisms in the organic infusion and that their generation thus depended on the germs contained in the air.

After an exchange of letters with Pouchet in April 1861 [17*, 18, 18*, 19], Pasteur reported the results of his experiments in a paper entitled *Mémoire sur les corpuscules organisés qui existent dans l'atmosphère* (1922, 210–320; Plate 2). He delivered a lecture on the

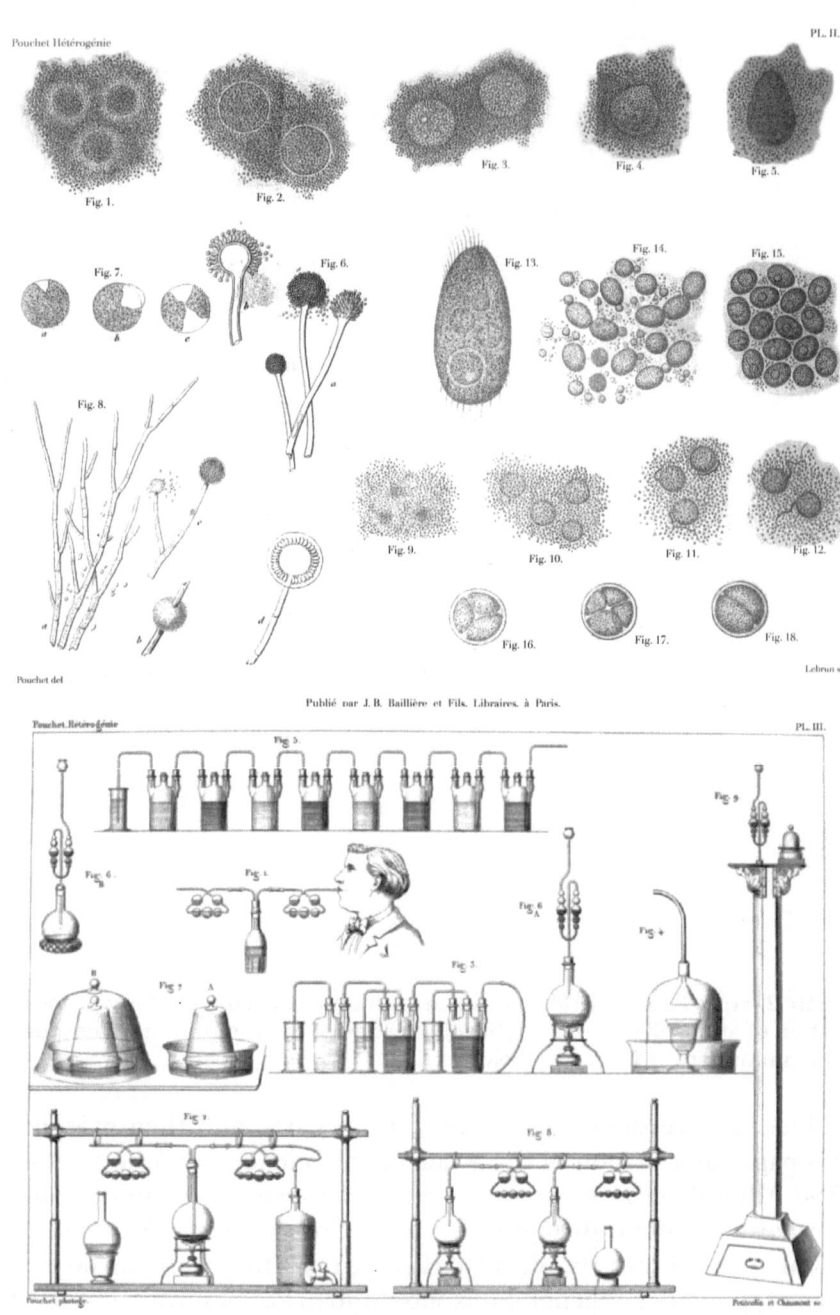

Plate 1. From Pouchet's *Hétérogénie:* microscope study of *Infusoria* (above), including *Aspergillus Pouchetii* (see Fig. 8); variants of the Schultze's apparatus used to produce the *Infusoria* (below).

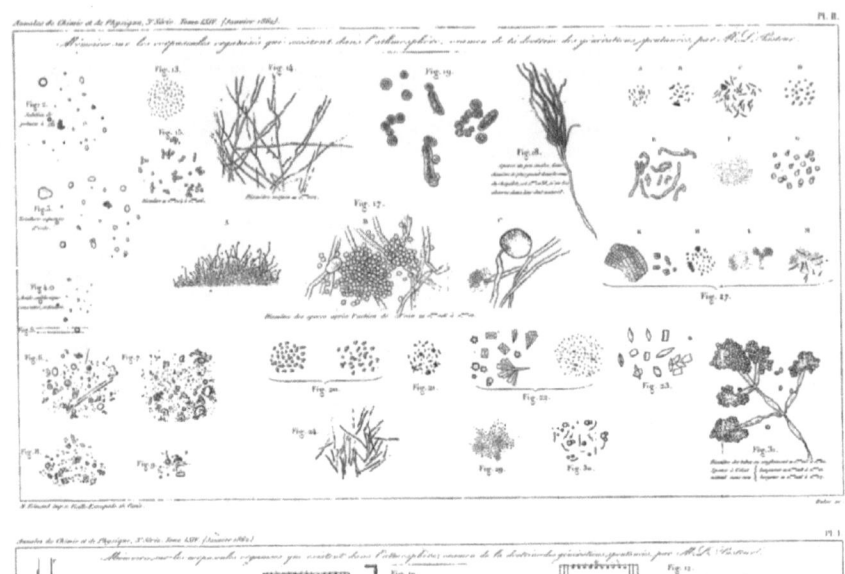

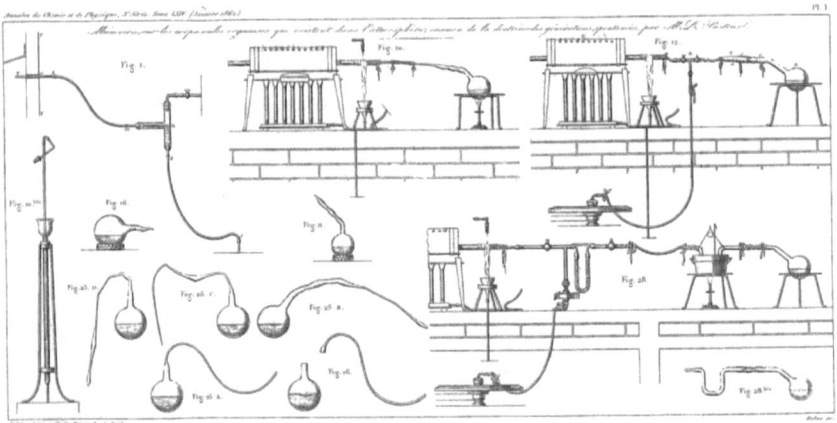

Plate 2. From Pasteur's *Mémoire* for the Alhumbert's prize: microscopic observation of atmospheric germs (above); the device used to prevent atmospheric contamination of the flasks (below).

subject before the Chemical Society of Paris, and then presented his paper at a session of the Academy of Sciences on June 15, 1862. The Alhumbert Prize commission, which was made up of Claude Bernard, Adolphe Brongniart, Victor Coste, Pierre Flourens, and Henri Milne-Edwards,[1] decided that the prize for resolving the problem of spontaneous generations should go to Pasteur in a ceremony to be held on December 29, 1862. Pouchet had submitted his own results regarding spontaneous generation to the commission on September

22 (BnF, naf 18111: 116), but withdrew from the competition after an interview with Milne-Edwards, who told him that he would be voting in support of Pasteur, ". . . because I saw his experiments and they entirely convinced me" (Pouchet 1864, xiv).

This award apparently marked the victory of Pasteur and the resolution of the controversy. But Pouchet, who had in the meantime allied himself with Nicolas Joly and Charles Musset in Toulouse, was not prepared to admit defeat. The three scientists undertook to repeat Pasteur's experiments, taking flasks containing organic material up to an altitude of 3,000 meters (9,800 feet) in the Pyrenee Mountains. They exposed four flasks to the air in the village of Rencluse (A, B, C, D), and four others on the glacier of Maladeta (E, F, G, H). The results, which they published in a note in *Comptes rendus de l'Académie des Sciences* in 1863, absolutely supported the theory of spontaneous generation. However, they only reported the data on four flasks (A, D, E, F) and Pasteur writes to his adversary on October 30th, "Would it be indiscrete of me to beg you to inform me as to what occurred in flasks B, C, G, H, of which your note makes no mention?" [38*]. Pouchet asked Joly to respond, suggesting that he attribute the omission of the results for these four flasks to a "faulty transition inadvertently suppressed by Mr. Musset" [38**]. Little satisfied with this explanation, Pasteur shared his reservations with the Academy of Sciences, setting off a discussion that would continue from September 21, to November 16, 1863. At the end of November *Pouchet* [39, 46*] demanded that the academy nominate a commission to adjudicate on the question once and for all. A commission was formed on January 4, 1864, consisting of Antoine Balard, Brongniart, Jean-Baptiste Dumas, Flourens and Milne-Edwards, which decided that both sides should repeat their experiments, setting a deadline of the month of March. The heterogenists appeared to be in full accord with this decision:

> We willingly take up the gauntlet that our learned adversary has thrown down, and we promise to conform, even more scrupulously than we have thus far, to each of the minute precautions that he has indicated as being rigorously indispensable. If a single one of our flasks remains unaltered after exposure to the air in Toulouse, we will candidly admit our defeat (Pouchet 1863, 845).

In the end Pouchet and his collaborators did not keep their promise to respect the conditions set by the commission, arguing first that their results would be compromised by the frigid temperatures of early spring. Pasteur suggested that the temperature during incubation might be artificially raised using a heating apparatus, but the commission instead decided to extend the deadline and accept reports submitted as late as the month of June, *against* the advice of Pasteur [39, 40]. In June Pouchet's team refused once again to conduct its experiments in front of the members of the commission. In a letter dated June 18, 1864, the academy offered the partisans of spontaneous generation yet another opportunity to repeat their experiments. But judging that the most favorable season for microbial activity had passed, Pouchet and his collaborators refused one last time [46*]. The commission therefore was only able to witness the experiments of Pasteur and it concluded that the results coincided exactly with the scientist's report of February 20, 1865 (Pasteur 1922, 637–647).

Pasteur did not wait for a final concession on the part of Pouchet to proclaim the demise of heterogeny. In a lecture delivered at the Sorbonne on April 7, 1864, he explained where his contradictors had erred and the experimental method that should be followed in order to avoid repeating their mistakes. This, in broad outlines, is how the controversy over spontaneous generation unfolded in the "constitutive forum." Perhaps the most striking feature of the story is that it can be divided into two distinct periods. While in the first phase (from February 1859, to June 1862) Pasteur found himself in the position of the experimentalist seeking to defend a novel thesis and Pouchet was the expert critic challenging his results, in the second phase (from June 1862, to June 1864) their roles were reversed. The two phases were separated by a crucial event—the bestowing of the Alhumbert Prize on Pasteur.

2. An Inventory of Asymmetries

When analyzing the resolution of a controversy according to the canons of SSK, it is necessary to lay aside the presupposition that there is a difference between the true and the false. One must postulate that there is a perfect symmetry between the two protagonists and their positions, and interpret the settlement of the debate as the result of *accumulated asymmetries* that may consist not only of scientific proofs but also such factors as personal prestige and political support. This form of radical constructivism does not allow one to take into account the error or truth of a scientific hypothesis, as this would involve making a retrospective judgment on scientific facts. From the perspective of SSK, truths in themselves do not exist, only pronouncements that may be more or less consensual. One task of the sociologist is to describe the stages in the construction of this consensus *without passing judgment on its content.*

1. Ambiguous Asymmetries

1. Parisian versus provincial. The sociologist Bruno Latour (1989, 438) viewed the debate as opposing a scientist living in the provinces far from the intellectual circles of the capital, and a Parisian who made use of his influential connections to ensure that the debate would be resolved in his favor. Latour here is following Pouchet's own version of events; for example, the scientist from Rouen repeatedly complained of the "Parisianism" that he came up against, writing to Joly, "Your origins and mine constitute an indelible sin" [34], and accusing the government of "piling all of its favors on the little corner of Paris alone" [36]. In 1862 Pouchet describes Milne-Edwards' intention to vote for Pasteur as a decision that "struck the entire province with ostracism in one fell blow" (1864: xiii).

Is this perceived asymmetry justified? One may argue that it is not. As Crosland has shown (1992), in this period the Academy of Sciences included many members who were considered to be "Parisians," but who were actually born in the provinces. Balard came from Montpellier (Hérault), Claude Bernard from Saint-Julien (Rhône), Chevreul from Angers (Maine-et-Loire), Coste from Castries (Hérault), Dumas from Alès (Gard), Flourens from Béziers (Hérault), Geoffroy Saint-Hilaire from Étampes (Essonne), Milne-Edwards from Bruges (Flandres), and Quatrefages from Valleraugue (Gard). Exactly how, and even more importantly why, would these "Parisians" have formed a united block against Pouchet? Furthermore, it is not precisely accurate to describe the naturalist as "Rouenese" per se. Pouchet had connections in Paris through his father, Louis-Ézéchias, who was personally acquainted with Lagrange, Haüy, Coulomb, and Laplace (Cantor 1992). Early in his career Félix-Archimède Pouchet obtained a post as *préparateur* (assistant) at the Paris Museum of Natural History through the good offices of Henri de Blainville (Cantor 1992) and he later helped to found, with the zoologist Isidore Geoffroy Saint-Hilaire, the Imperial Zoological Acclimatisation Society of Paris. The naturalist was on close terms with many members of the Academy of Sciences. Finally, it may be added that the Minister for Education Gustave Roulland, to whom he appealed for support in 1863 during the course of the controversy, like him came from Rouen [27]. The picture of an elite milieu in the capital that was hostile *a priori* to a scientist from the provinces is a projection with no empirical foundation.

The notion that *le tout Paris* had been bought off by Pasteur is just as much of a caricature. At the start of the controversy, this son of a tanner, born in Dole in the region of Franche-Comté, had recently left his post as dean of the Faculty of Sciences at the University of Lille and moved to Paris, where he had been appointed director of scientific studies at the École normale supérieure (1857). Even if this provided Pasteur with the opportunity to form ties with many Parisian intellectuals, it would be questionable to assume that he owed his scientific achievements to these connections. Such a proposition could be defended only if one were able to establish: (i) That the fact of his residence in Paris increased the number of his *effectual contacts* with scientists there; (ii) That these contacts were not competitive or adversarial; and (iii) That he would have allowed the scientific content of his work to be influenced by these exchanges. Any such chain of circumstances is far from being substantiated. Pasteur, who is reputed to have been quite taciturn by nature, did not frequent the intellectual salons of the period (Cantor 1992). Furthermore, it is known that he had antagonistic relationships with many of his colleagues. At the same time, it is a fact that researchers can remain on courteous terms while holding divergent views. The opposition "Paris *versus* Rouen" therefore has no intrinsic significance.

2. *Corresponding member versus full member of the French academy of sciences.* Much stress has been laid on the differing status of the two scientists within the *French Academy of Sciences*. It is obvious that a full member would enjoy a position of greater scientific authority than a corresponding member. But nothing has been said about the *dates* on which the two scientists were admitted to the academy. Pouchet was elected on December 17, 1849, in recognition of his work in zoology. Pasteur would be admitted much later—on December 2, 1862, to replace the (recently deceased) Henri Hureau de Sénarmont in the Section of Mineralogy—following his contributions in the areas of crystallography and rotatory polarization. Here it is necessary to take into account a double asymmetry.

Pasteur had already *completed* his experiments on spontaneous generation when he was elected to the Academy of Sciences in December 1862, and he had submitted his *mémoire* to the secretary of the academy on 15 June 1862 (Pasteur 1951, 109). This chronology undermines the relativist interpretation of events, according to which the commissions convened in 1862 and 1864 "were only composed of colleagues of Pasteur who were almost all convinced in advance" that his hypothesis was correct (Latour 1989, 435). Collins and Pinch present the SSK view: "Pasteur's most decisive victory—his defeat of fellow Frenchman Felix Pouchet, a respected naturalist from Rouen, in front of a commission set up by the French Academy of the Sciences—rested on the biases of the members" (Collins and Pinch 1993, 80) and Vinck repeats the argument without providing any corroboration: "[The Academy of Sciences] instituted two commissions composed of members who were in favor of Pasteur; they accorded him unconditional support and had nothing but disdain for Pouchet" (1995, 114). In fact it is difficult to see how Pasteur could have "relied solely on the support of his friends in the academy"—at least before December 1862. What is more, a review of Pasteur's correspondence shows that his only close ties among the academicians were with his former professor and mentor Antoine Balard, and with the chemist Jean-Baptiste Dumas and the physicist Jean-Baptiste Biot (who passed away, however, on February 2, 1862, without having taken part in the discussions of the commission for the Alhumbert prize). Pasteur was acquainted with Pierre Flourens, Michel-Eugène Chevreul, Henri Milne-Edwards, Augustin Serres, and Claude Bernard but his intercourse with them remained on a strictly formal footing, as can be seen from a perusal of his letters.[2]

As for Pouchet, due to his long association as a corresponding member (1849–59) and his scientific activities, he was well known at the institute. Twenty-five notes in his name had already appeared in the *Comptes rendus de l'Académie des Sciences*, well before the competition for the Alhumbert prize in 1862, whereas up to that year Pasteur had only published twelve notes. Pouchet enjoyed relationships of long standing with numerous members of the academy. His scientific doctrines had been profoundly influenced by the ideas of Henri-Marie Ducrotay de Blainville and Étienne Geoffroy Saint-Hilaire, both of whom he had met while working at the Paris Museum of Natural History (Cantor 1992) and would continue to frequent afterwards. He maintained friendly relations with the latter's son, Isidore Geoffroy Saint-Hilaire, who accorded his papers on heterogeny a favorable reception when they were published [7, 17, 20].

Pouchet's ties with key members of the academy are worth emphasizing. He had close relations with Victor Coste, who was nominated to the commission on spontaneous generation, due to their shared interest in pisciculture. In a letter in which he suggested that they might collaborate together on some research [5], Coste showered him with "*mille compliments affectueux* [a thousand affectionate compliments]", referred to their *"vieilles amitiés* [friendship of long date]", and closed with the warm salutation *"tout à vous de coeur* [to you with all heartfelt sentiments]" (Rouen Museum of Natural History, FAP 3047). Pouchet's links to Pierre Flourens—the Permanent Secretary of the Academy of Sciences and a member of the commission of experts—were no less close. He had known Flourens since 1843 and would recall watching his son Gustave grow up [49]. A letter written by Flourens in 1847 contains very flattering references to Pouchet's work *Théorie positive de l'ovulation* [1, 2, 3], and closes with the conveyance of *"toute son affection* [all his affection]" [4]. Pouchet dedicated *Hétérogénie* to another member of the institute whose esteem he enjoyed—Pierre Rayer [50]—"in public homage of his gratitude" (1859, x). Finally, when recalling the fraught closing months of the controversy, Pouchet retained an emotion-filled memory of Serres who, "with an obligingness that I will never forget, placed his laboratory at my disposition" (1864, xv).

Pouchet was therefore anything but an obscure scientist from the provinces to such prominent figures as Coste, Flourens, Geoffroy Saint-Hilaire, Rayer, and Serres, with whom he corresponded regularly. The SSK case for accumulated asymmetries thus rests on a tenacious, but mistaken presupposition, as had already been identified by Crosland (1992): the French Academy of Sciences was not a homogenous and unified body.[3] A detailed study of the institution shows that its membership has always been characterized by a wide diversity in terms of its social origin, age (the average age of newly elected members was 42 years), and political opinion (Crosland 1992, 101, 177, 189).

3. Close versus distant ties with the emperor. Pasteur has often been described as being close to the emperor "during the decade of the 1860s" (Latour 1989, 442; Mendelsohn 1987, 111), without, however, any documentation being provided of the dates, circumstances, and precise nature of these ties. It has been assumed that Pasteur received subsidies for his research and concrete support for his ideas through his privileged relationship with the emperor. Collins and Pinch carry the argument even further, affirming: "The oppositions [to Pasteur] were crushed by political maneuvering, by ridicule, and by Pasteur drawing farmers, brewers, and doctors to his cause. [. . .] As in so many other scientific controversies, it was neither facts nor reason, but death and weight of numbers that defeated the minority view; facts and reasons, as always, were ambiguous" (Collins and Pinch 1993, 80).

Let us examine these relations in the circles surrounding Napoléon III more closely. Since the École normale, of which he had been appointed administrator and director of studies, was not a research institute, Pasteur was forced to ask for supplementary grants in order to pursue his work. One year after he took up his post, records show that the Minister for Public Instruction agreed to cover the costs of the *maintenance* of the buildings, nothing more. His own applications having met with only limited success between 1859 and 1860, Pasteur decided to ask the director of the École normale and Colonel Favé, the emperor's personal orderly, to intercede for him. In this he was not seeking support for his scientific theories, but the means to keep his laboratory running. As he writes in February 1862, to Favé: "In seeking to bring this work to the attention of the Sovereign, I harbor the covert desire to acquire the resources to pursue it with greater liberty. This small laboratory [. . .] is no longer sufficient for my research projects" (1951, 99).

A simple examination of the chronology shows that this process could not have had any consequences in the debate over spontaneous generation as there was no question of receiving funding for specific experiments nor any attempt to bring about a political settlement to the controversy.

Primo, the decision to enlarge the laboratory at the École normale supérieure was approved on May 31, 1862, by which time Pasteur had already submitted his *Mémoire sur les corpuscules organisés*.

Secundo, there are documents proving that neither Favé nor the emperor became involved in the debate over spontaneous genera-tion. The following incident testifies to this. Colonel Favé, who was a graduate of the École Polytechnique, aspired to a seat in the Academy of Sciences and sought the support of Pasteur for his candidacy. Even though the debate in which he was embroiled was at its height, in April 1864, Pasteur sends him the forthright response [39*]:

> I was far from suspecting, Sir, that if I one day had the satisfaction of learning that you were a candidate for a vacant seat at the Académie, I would be among those who would fail to contribute to your success. This however is the acute regret that is reserved for me. I do not see how I can avoid voting for M. Foucault.

The physicist Jean-Bernard-Léon Foucault (1819–68) was indeed elected a few months later, and Favé would have to wait until 1876 before his application was granted a more favorable reception. Pasteur's reply was dated April 3, 1864, and therefore *well before the decision handed down by the commission of the Academy of Sciences*. If Pasteur had felt that he was in some way beholden to Favé, then one might presume that: 1) he would have voted for him, or if he did not, 2) then Favé, moved by a sense of resentment, might have sought between April and June to convince the commission to pass judgment against the scientist. Clearly, neither of these eventualities transpired, and it may be concluded that Napoleon III and his orderly avoided any involvement in his dispute with Pouchet.

But was Pouchet himself really so isolated from the centers of power as is claimed? It is pertinent to note here that the naturalist had been an intimate friend of Prince Charles-Lucien Bonaparte, a first cousin of Napoleon III who settled in Paris in 1850 and was himself a member of the Academy of Sciences (Cantor 1992). Although Charles died in 1857 and therefore could not have served as an intermediary between Pouchet and the emperor in this controversy, the scientist certainly might have exploited his friendship in other ways. And there were further avenues open to Pouchet. Lectures organized by the Société impériale d'acclimatation, of which he was cofounder, were regularly attended by the prince and government ministers [16]. In fact Pouchet never hesitated to importune powerful political figures and seek their support in his scientific rivalry. At his instigation Joly writes on February 8, 1862, to the Minister of Education Roulland, urging him to "*tout remuer* (stir everything up)" in Paris [23]. In March Pouchet attempts to persuade the minister to intercede in his dispute with Milne-Edwards about spontaneous generation [24]. In May 1864, Pouchet writes to Roulland once again, this time claiming to be in an unassailable position with regard to the question of spontaneous generation vis-à-vis Pasteur, "who owed his fame to him" [38]. A month later Pouchet managed to meet with the Dean of the Faculty of Paris, following which he writes to Noël with great disappointment that his plan "did not work," although it is not possible to determine exactly what the nature of this "plan" was [45]. In the end, to the immense despair of the scientist from Rouen [44], the political figures to whom he appealed refrained from taking part in the settlement of the controversy.

4. Researchers acting in good or bad faith. In his analysis of the text of the lecture delivered by Pasteur at the Sorbonne on April 7, 1864, Latour maintains that the scientist deliberately sought to associate his rival's position with atheism as part of his campaign to defeat him. Latour interpreted this as a reversal of position on the part of Pasteur, a "maneuver" which demonstrated that he was acting in "bad faith"

(1989, 433). With regard to any possible associations between science and spirituality, the scientist explicitly stated in his lecture, "Here it is neither religion, nor philosophy, nor atheism, nor materialism, nor spiritualism that stand. [. . .] Here it is a question of fact" (1922, 345). As far as the accusation of bad faith on the part of Pasteur is concerned, an analysis of the letters of Pasteur and those of Pouchet is sufficient to dismantle this supposed asymmetry.

Pasteur probably never accepted Pouchet's results, but he did acknowledge his opponent's efforts to test his theory experimentally. In a letter dated April 3, 1861, Pasteur writes: "The difference in our opinions on the celebrated question of spontaneous generations does not prevent me from appreciating [. . .] the laudable efforts that you have made to introduce experimentation into questions of natural history" (1951, 84). Even in private Pasteur always spoke of his adversary with a measure of restraint:

> I am continually being contradicted by two naturalists [. . .]. But I do not waste time responding to them. Let them say what they like. I believe I know what the truth is. They do not know how to conduct a proper experiment. This is not an easy art. One must bring to it, in addition to certain innate abilities, long experience that naturalists in our day generally lack (1951, 72).

Pasteur did not seek to enter into a polemic over what he regarded as the fantasies of the naturalist and refrained from using the press to discredit his contradictor.

A symmetrical analysis of the correspondence of Pouchet shows that, in contrast, the director of the Rouen Museum of Natural History was prepared to use any arms that came to hand—whether fair or unfair—to attack his opponent. Thus, when Pasteur wrote to Pouchet to enquire as what had happened to the "missing flasks" in the Maladeta experiment [38*], Pouchet sidestepped this question and asked Joly to respond in his place with the tactical instruction: "You can see where he wants to arrive. Weigh your phrases well . . ." [38**].

When referring to his enemy in his letters, Pouchet used a number of expressions designed to place him in a bad light. In front of his friends he referred to Pasteur slightingly as that "Paracelsus II" who was always seeking to contradict him (BnF, naf 18111: 76, 115, 125), a "charlatan," a "joker," or a "visionary" [27] who "had gone astray" [12] and managed to produce nothing but "muddled ranting" [21] and "ridiculous experiments" [27]. He demanded that his friends combat Pasteur's "outdated chemical notions" on his behalf [26].

The tone of these comments is echoed in the campaigns of denigration launched by various journalists, including his friend Noël [12, 25, 37, 43, 47]. The archives of the Rouen Museum of Natural History [FAP 52] contain many letters written by Pouchet to, among others, Castelnau of the *Moniteur des Sciences*, Legrain at *Progrès par la Science*, Meunier of the *Courrier des Sciences*, Abbot Moigno at *Cosmos*, Paschal at *Époque*, Piton-Bressaut of the *Ami des Sciences*, and Quesneville at the *Moniteur scientifique* and they all resonate with the same tone.

It was therefore Pouchet, and not Pasteur, who bears entire responsibility for blurring the boundaries in this controversy between the constitutive and the unofficial forums. This asymmetry contradicts the hypothesis that the "bad faith" was on Pasteur's side. It is in fact difficult to view the dogmatism [30, 32, 46], the inappropriate use of the organs of the press, and the falsification of documents (as we will see below) [7] by Pouchet as evidence of a high standard of ethical conduct.

2. Hidden Asymmetries

In his study, Latour (1989) took it upon himself to highlight certain asymmetries in the debate over spontaneous generation. We might expect an essay delineating a program to be written in a somewhat summary form, but a perusal of the details of the controversy shows that his catalogue of asymmetries is far from exhaustive. In reality Latour's thesis—that the settlement of the controversy was the result of an accumulation of asymmetries—is based on his own *drastic selection* among the various possible asymmetries. There are many others that lead to precisely the opposite conclusion.

1′. Professional versus dilettante. The first overlooked asymmetry pertains to the professional status of the two researchers. Unlike Pouchet—who was the director of the Rouen Museum of Natural History at the time of the dispute—Pasteur did not conduct his research under the auspices of a research institute. He was the *"directeur des études* (director of studies)" at the École normale. This was an administrative position consisting of "the supervision of the financial and sanitary arrangements, maintaining general discipline and relations with the families and the scientific and literary establishments frequented by the students." He served in this position from 1857 to 1867 and conducted his experiments during his spare time in a makeshift laboratory set up in the attics of the École normale. On June 28, 1858, Pasteur writes to Minister Roulland to request that a proper laboratory be installed. He received some meager funds, with the proviso that they not be used for "the conduct of [research] work that would compromise the convenience of the persons housed in these buildings" (Pasteur 1951, 28). In December 1859, Pasteur requested another subsidy (1951, 60), but the minister refused to renew what he considered to be a "special allocation." Although Pasteur submitted further applications in 1860 and 1861, the enlargement of the laboratory would not be approved until 1862. Pasteur describes his situation in a letter to Favé in 1863:

> It is precisely during this crucial period in my career that I have found myself charged with administrative duties that deprive me of the right to have a laboratory and funding for the courses and experimentation that all the professors who are actively engaged in teaching enjoy as a matter of course (1951, 148).

Pasteur finally received official permission to expand his laboratory at the École normale on May 31, 1862. Having already completed his studies on spontaneous generation, it can be stated with certainty that his experiments were conducted in the absence of any institutional support. What is more, any statements or publications by him would have been those of a "functionary" of the École normale, whereas Pouchet was speaking as the director of an important science museum. This asymmetry in professional status partially explains the indignant reaction of the scientist from Rouen to Pasteur's challenge to his authority.

2'. Well-established researcher versus young aspiring researcher. An analysis of the controversy cannot ignore without comment the relative ages of the two protagonists. When the debate began, Pasteur, who was only 37, was daring to challenge a senior scientist, the 60-year-old Pouchet. Farley and Geison (1991) and Latour (1989) mention this difference, but do not discuss its sociological implications. At the time of the debate, Pasteur had not yet achieved the eminent reputation in the fields of medicine and biology that he has today. In 1860 he was only known for his research on rotatory polarization in crystals and molecular dissymmetry. His sole published work in the field of biology was a study on *Penicillium* (1860). The groundbreaking discoveries on pasteurization (1867), the microbes responsible for infection (1870), and the vaccination against rabies (1885) for which he became celebrated still lay in the future. Therefore, the very principle of asymmetry would require that one eliminate from the analysis of this controversy the imagined consequences of a debate conducted between the great scientist (Pasteur) and the obscure provincial (Pouchet), because it is certain that during the period of the controversy Félix-Archimède Pouchet had a much more illustrious reputation than his younger colleague, particularly in the field of biology where Pasteur was still a novice.

The difference in age between Pouchet and Pasteur does have a sociological significance. The analyses of Merton and Zuckerman (1973, 537–559) on the gerontocracy that prevails in the sciences illuminates what seems to be the surprisingly discourteous behavior of Pouchet. The system of "peer review," while egalitarian in principle, in reality tends to reward the more experienced researcher, who will—*de facto*—be older in age.

At the same time, one may note the rapid upward trajectory of Pasteur's career during the course of the controversy. In 1859 he challenged the work of his more senior colleague in the latter's own territory, and during each of the next three years he was awarded a prize by the French Academy of Sciences (the Montyon, Zecker and Alhumbert awards). Pasteur was elected a full member of the academy in 1862, while Pouchet's position as a corresponding member remained unchanged after his election in 1849. This difference in their career paths is sufficient to explain the resentment of Pouchet, who could not ignore the brilliant rise of his younger colleague, and it is understandable that personal rancor and jealousy would have colored his analysis of purely scientific questions.

3′. Prolific versus modestly productive researcher. A systematic study of the asymmetries in this controversy could also focus on the scientific productivity of the two researchers. The list (Appendix 1) of their publications before June 1864 (the date on which the experiments of Pasteur were officially recognized by the Academy of Sciences) allows us to measure and compare the productivity of the two scientists. It suffices to compile a histogram of the publications that appeared between 1830 and 1865 (here divided into five-year intervals) (Figure 1). This immediately reveals a flagrant asymmetry between the two researchers. In 1864 Pasteur and Pouchet had a total of thirty-five and seventy-six publications to their names, respectively. Limiting our analysis to the period 1859–64, the productivity of Pasteur was still inferior to that of Pouchet (twenty-four vs. thirty-four papers). This shows manifestly that *scientific productivity does not constitute a predictor of scientific success.* The differentiation Cantor (1992) proposes between research papers intended for a specialist audience and articles written for the general public does not resolve the

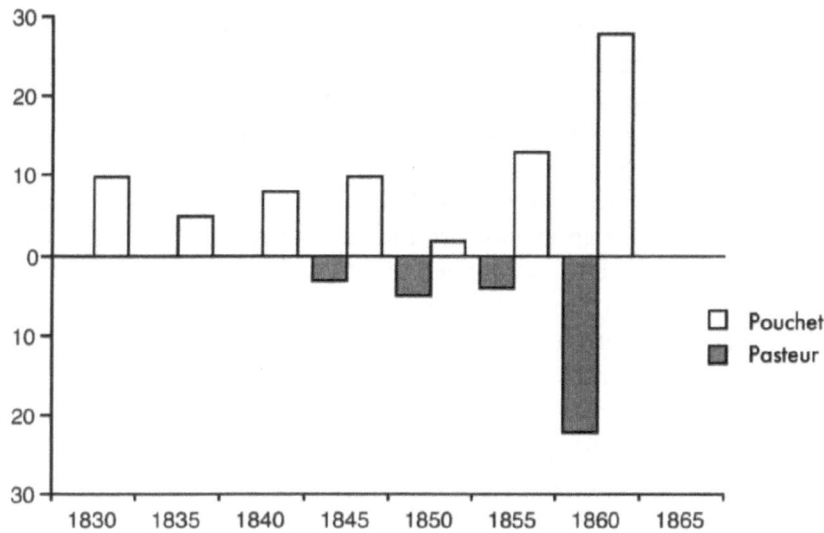

Figure 1. The papers published by Pasteur and Pouchet (1830–1865).

problem: it is not always a simple matter to determine, particularly in the case of Pouchet, whether a text falls into one domain or the other, or both.

One reason why it is difficult to draw any conclusions from a comparison of the number of works published is that Pouchet and Pasteur had very different notions of what constituted a scientific publication. Pasteur preferred to communicate through succinct notes whereas Pouchet was an expansive "elaborator." This explains why the latter's book *Hétérogénie* came to 704 pages. But of what was this treatise on spontaneous generation actually composed? One finds a preamble (thirty-two pages), a chapter on metaphysics (forty-three pages) and, after an "experimental refutation," a histological study of *infusoria* (134 pages) followed by "geological proofs" (211 pages). The question of spontaneous generation *stricto sensu* takes up no more than 26 percent of the whole (1859, 138–325). The preface to *Hétérogénie* states somewhat disingenuously: "The totality of this work can naturally be divided into two sections: the experimental part, which is the most fundamental, and the theoretical part, which only forms an accessory fragment" (1859, vii), in what appears to be a rhetorical notification rather than a factual description of the book.

4'. Scientific style: demonstration versus illustration. A detailed analysis of the scientific texts of the two protagonists in this controversy highlights yet another asymmetry which pertains to their respective writing styles. The papers by Pasteur are dense and carefully constructed on the basis of experimental evidence. The writings of Pouchet abound in prolix commentary and rely on illustration, in which arguments drawn from authoritative sources are used to support, or compensate for the lacunae in his experiments. Hence Pouchet's frequent references to Geoffroy Saint-Hilaire.[4]

This trait, which was noted by Roll-Hansen (1979, 277) and Gálvez (1988, 353), exposes an even more radical contrast between the two men in terms of their "styles of reasoning" (Hacking 1992).[5] Schweber (1997), who applied this notion to a study of statistical knowledge in the nineteenth century, shows that the disagreement between the participants in a controversy may stem from cognitive factors. The sheer length of Pouchet's *Hétérogénie* and some of his letters [18, 19] show that for the naturalist the question of spontaneous generation was inseparable from certain considerations that Pasteur *did not consider to form part of the same problem*. Pouchet writes: "I will show to whomever wishes, examples of yeast that is born spontaneously, yeast that is germinating, yeast that is beginning to sprout, yeast that is fructifying [. . .]" [19]. But Pasteur only contested the *first* of these four assertions, ignoring the others. This is an indication that the thesis of heterogeny did not have the same meaning for the two researchers, because their approach to the problem was shaped by their different scientific interests. Whatever the biological interest of a cytological study on the development of microorganisms, this approach could not address the question of their *origin*. Pouchet committed two errors simultaneously, first in believing that his facts constituted proof of the way in which microorganisms appear and secondly in viewing—unlike Geoffroy Saint-Hilaire [17]—Pasteur's research as the intrusion by a novice into an area in which Pouchet was the recognized expert [11, 12].

Their diametrically opposed styles of reasoning (demonstration *versus* illustration) can be viewed in relation to the scientific training of the two researchers. Pouchet was a naturalist while Pasteur was a chemist. Pouchet had spent several periods abroad, in Switzerland and Germany (Cantor 1992, 57), and had embraced the notions of *Naturphilosophie* and German Romanticism. In contrast, the works of Pasteur exhibit a pronounced analytical character. As a consequence there was little likelihood that the two men would develop a similar approach to microbiology. Their emotional stake in the controversy was also completely different. It was the chemist Pasteur who contested the lack of rigor in the naturalist Pouchet's experiments. It was then the naturalist who became incensed, demanded that his collaborators cross swords with the enemy, and then disappeared from view. This correlation between a scientist's style of reasoning, his professional training, and the type of social behavior that he engages in should not surprise us. In his studies of how modern research laboratories function, Terry Shinn (1980) observed a close correspondence between the training received by a scientist and the mode of social behavior that he tended to adopt subsequently. The contexts are probably too different for one to extrapolate Shinn's results to the controversy between Pasteur and Pouchet, but a comparison does shed light on the links that may justifiably be sought for on the basis of "styles of reasoning."

5'. The skeptical versus the committed attitude. Along the same lines as the "styles of reasoning" just discussed, one can compare the attitudes of Pasteur and Pouchet to scientific activity in general. A study of their correspondence reveals that on many occasions Pasteur's detached skepticism contrasted with the passionate involvement of his adversary from Rouen. At the debut of the controversy Pasteur writes to Pouchet, "In my opinion, Sir, the question is entirely and completely devoid of any decisive proofs" (1951, 46). This prudent attitude had nothing in common with the stubborn convictions that continually make themselves felt in Pouchet's *Hétérogénie*:

> When upon meditation it became evident to me that spontaneous generation was one of the means employed by nature for the reproduction of beings, I applied myself to discover by which procedures one might arrive at highlighting these phenomena [. . .]
> As for us, we are fighting under a most respectable and most imposing banner [a list follows of sixty-one ancient and modern authors]
> We, we are fighting with a more magnanimous and more disciplined army [. . .] we aspire to make our opinions prevail, confident in our strength [. . .]
> What one may reproach the adversaries of *spontéparité* [heterogeny] with is their having remained immobile in the face of the ascending progress of the philosophical sciences [. . .]
> Although I found myself rationally quite permeated with evidence of the thesis that I must defend, I will set to work to demonstrate it experimentally (1859, vi, 2, 5, 93 and 255 respectively).

Thus, while Pasteur remained skeptical with regard to all preconceived solutions, Pouchet was convinced *a priori* of the truth of the theory of heterogeny. His experiments reflect the consequences of this position. After having ascertained to his satisfaction that "air bubbled through sulfuric acid is incapable of spontaneous generation" (1859, 252), Pouchet repeated Schulze's experiments with the five-ball apparatus of Liebig, and reported nonetheless that despite this treatment spontaneous generation took place. On the twenty-fourth day he saw signs of *Penicillium* on the sides of the flask that was connected to the apparatus, while the flask left exposed to the air showed no signs of the blue mold (1859, 272). This was not the only occasion on which Pouchet reported a surprising experimental outcome.

The adherents to SSK tend to exaggerate Pouchet's gifts as an experimentalist.[6] In their view Pasteur was mistaken to dismiss his colleague's results out of hand, because the naturalist was working with infusions of hay that probably contained a microorganism—*Bacillus subtilis*, discovered by Cohn and Tyndall in 1876—which can resist temperatures of up to 100°C (Geison, 1981, 372; Latour 1989, 445; Farley and Geison 1991, 136). This argument is only partially admissible, however, for both historical and scientific reasons.

Primo, Pasteur had quite early on suggested to Pouchet, in a letter dated February 28, 1859, that he abandon his experiments with boiled hay infusions (1951, 45–46). In January 1861, Pasteur demonstrated that certain germs remain fertile in an atmosphere heated to a temperature just below 120°C, becoming sterile only when brought to a temperature of 130°C for twenty to thirty minutes.

Secundo, Pouchet dismisses Pasteur's hypothesis that some microorganisms can survive in boiling water as a "marvelous and enchanting" invention [14]. Notwithstanding this, the naturalist observed microorganisms in experiments conducted in air that had been heated to 150°C (Pouchet 1860). His disciple Pennetier even reports, ignoring the objection made by Milne-Edwards that a temperature of 100°C is not sufficient to kill certain hay germs, that Pouchet had successfully "brought decomposable matter to 200°C, 250°C, and higher, without preventing the spontaneous appearance of protoorganisms" (1907, 36). How was this possible, as no microorganism—not even *Bacillus subtilis*—can survive being heated to 120°C, the temperature at which cytoplasm coagulates? The SSK response is given by Roll-Hansen (1979, 290) for the experiments conducted on the Maladeta massif. Joly and Musset in fact discovered in their flasks some microorganisms that could not even resist a temperature lower than 100°C and therefore the two positive results reported were due to contamination of the flasks! This is why some sociological accounts of the debate never cease to surprise: "Pouchet constantly presented empirical proofs to support his positions in favor of the theory of spontaneous generation, which Pasteur often could only counter with his preconceived views" (Vinck 1995, 113).

For Pouchet, therefore, was the thesis of heterogeny more a matter of metaphysical conviction? There is convincing evidence for this in his correspondence. Pouchet felt that biology should follow "the upward movement of the philosophical sciences" and claimed for itself "the prize for dialectics" [21]. Guided by a rare spirit of modesty [10, 38], Pouchet believed totally in the strength of his position, without wavering from 1859 to 1864, a fact that reveals his dogmatic attachment to the theory of heterogeny, the very antithesis of the scientific ethos and its "organized skepticism" (Merton 1973). He writes to his collaborator on the faculty in Toulouse, "Our cause is definitely won," "we form a terrifying and indestructible phalanx," and "we are invincible" (BnF, naf 18111: 26, 30, 77). In the same vein: "I will never give up the battle" [30]; "I will not relinquish [the fight against Pasteur] until he is suffocating beneath the weight of the stones of heterogeny" [32]. Berating his allies each time they showed the least sign of indulgence toward the enemy [25], the certitude shown by Pouchet sometimes bordered on religious fanaticism. He writes to Joly, "Let us set aside the things of this earth [. . .] and speak only of the things of Sacred Science." The sole question at issue here—it would have been clearly understood—was that of heterogeny, the doctrine that he described elsewhere as a "Holy Cause" and a "Holy Religion." This is why Pouchet felt that there was an imperative need to combat the "Hydra of chemistry" (BnF, naf 18111: 109, 35, 46, 52, 140, 182). In 1862 he had already declared that Paris was nothing other than a "dreadful Babylon" [33] and Pasteur therefore had to "expiate" his sins [41]. Finally, in July 1864, all of this would lead Pouchet to declare, with remarkable lucidity, that "Pasteur has been crushed" (*sic*) [46].

6'. Upstanding researcher versus dishonest counterfeiter. One of the most delicate points in this case must still be examined. Should one consider that the differences in style of reasoning (demonstration *versus* illustration) and in attitude (skeptical *versus* committed) between the protagonists exercised a profound influence on the scientific debate and on the merits of the point being defended, or should one regard this influence as being confined to secondary aspects of the controversy? We have proof that Pouchet's dogmatism led him to commit acts that violated the most elementary principles

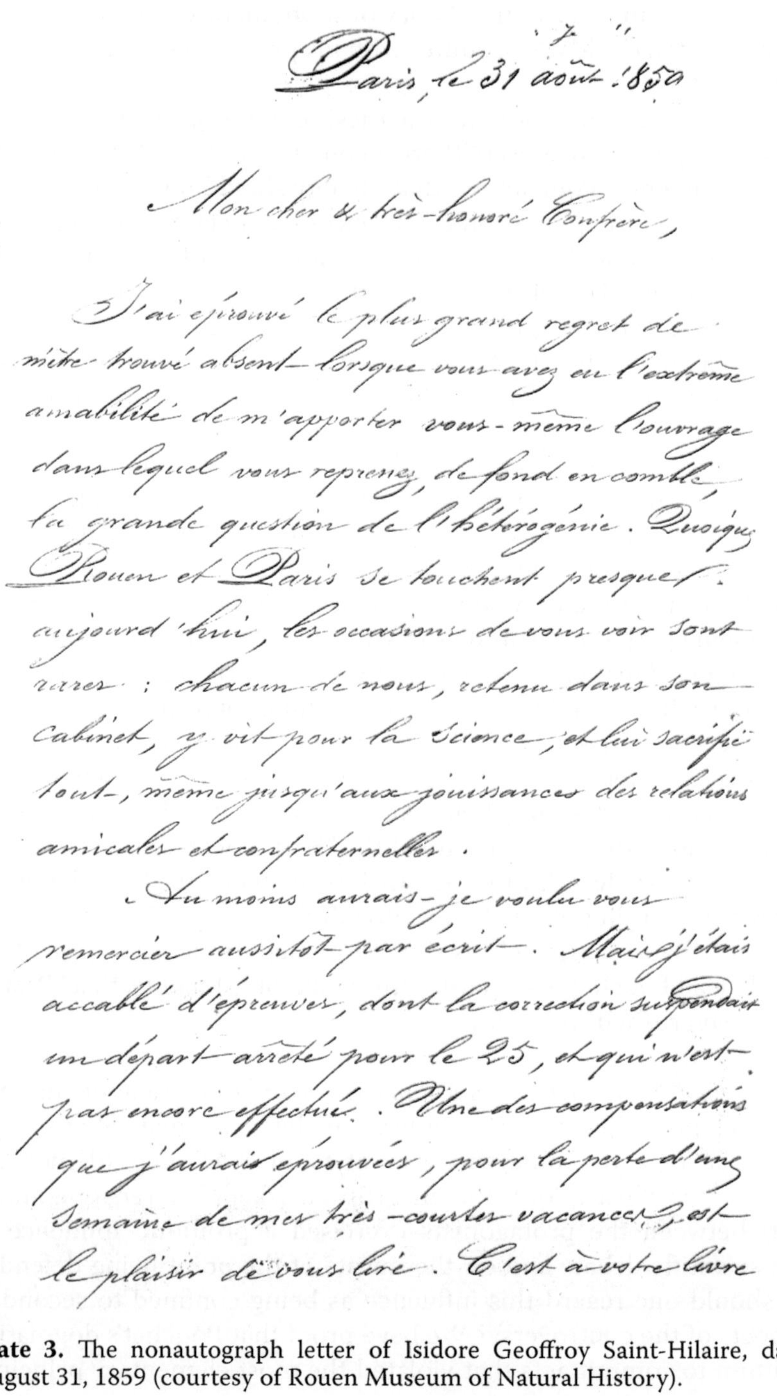

Paris, le 31 août 1859

Mon cher & très-honoré Confrère,

J'ai éprouvé le plus grand regret de m'être trouvé absent lorsque vous avez eu l'extrême amabilité de m'apporter vous-même l'ouvrage dans lequel vous reprenez, de fond en comble, la grande question de l'hétérogénie. Quoique Rouen et Paris se touchent presque aujourd'hui, les occasions de vous voir sont rares; chacun de nous, retenu dans son cabinet, y vit pour la Science; et lui sacrifie tout, même jusqu'aux jouissances des relations amicales et confraternelles.

Au moins aurais-je voulu vous remercier aussitôt par écrit. Mais j'étais accablé d'épreuves, dont la correction suspendait un départ arrêté pour le 25, et qui n'est pas encore effectué. Une des compensations que j'aurais éprouvées, pour la perte d'une semaine de mes très-courtes vacances, est le plaisir de vous lire. C'est à votre livre

Plate 3. The nonautograph letter of Isidore Geoffroy Saint-Hilaire, dated August 31, 1859 (courtesy of Rouen Museum of Natural History).

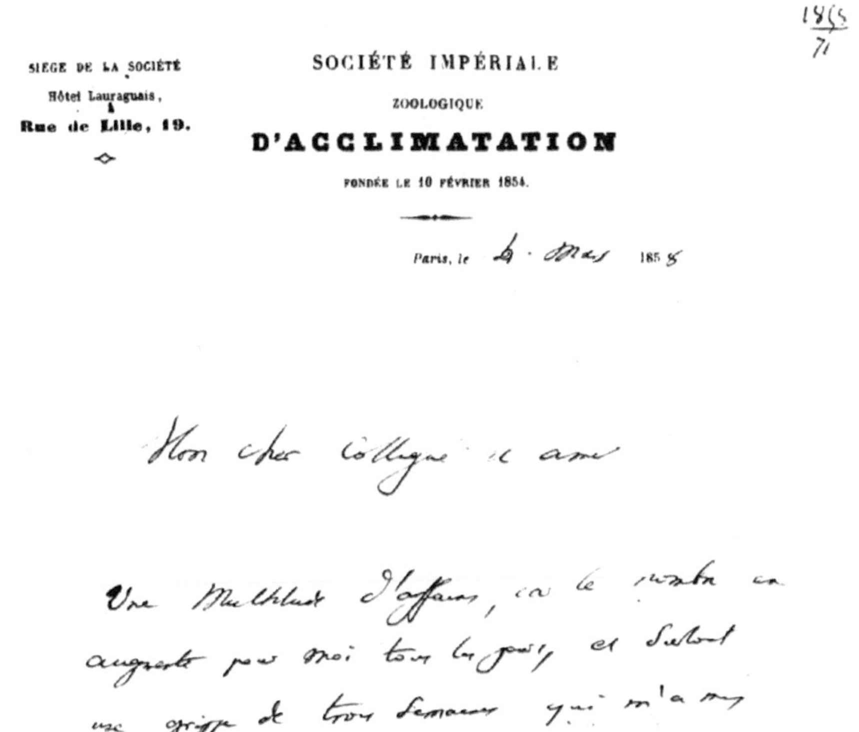

1865
71

SIEGE DE LA SOCIÉTÉ
Hôtel Lauraguais,
Rue de Lille, 19.

SOCIÉTÉ IMPÉRIALE
ZOOLOGIQUE
D'ACCLIMATATION
FONDÉE LE 10 FÉVRIER 1854.

Paris, le 4 Mars 1858

Mon cher Collègue et ami

Une multitude d'affaires, qui se semble va augmentant pour moi tous les jours, et surtout une grippe de trois semaines qui m'a mis en retard pour tout, m'ont empêché de vous répondre plus tôt.

J'ai entendu les ... de la thèse de votre fils. Je l'ai lue entièrement, et avec beaucoup d'intérêt ; car il n'est pas de questions qui m'intéressent plus vivement que celles qu'il traite. Je regrette que l'éloignement où

Plate 4. An autograph letter of Isidore Geoffroy Saint-Hilaire, dated March 4, 1858 (courtesy of Rouen Museum of Natural History).

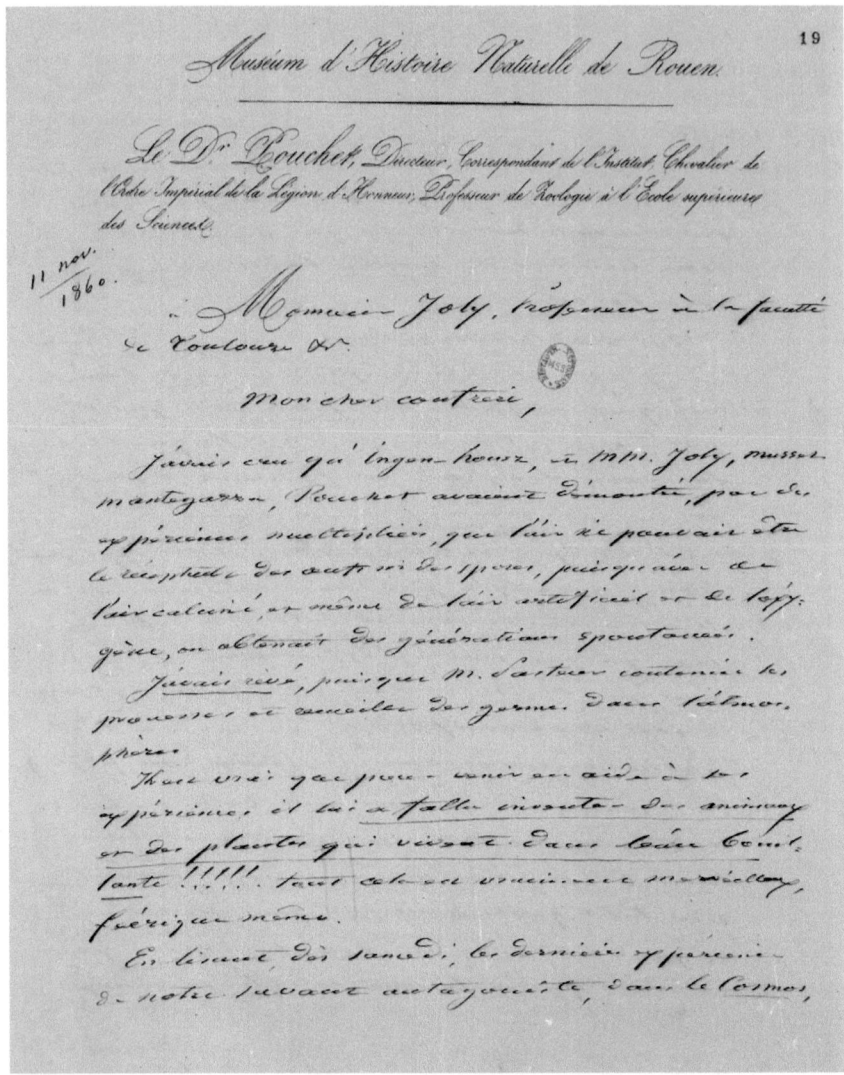

Plate 5. The letter of Pouchet sent to Joly on November 11, 1860 (courtesy of Rouen Museum of Natural History).

	Lettre d'Isidore Geoffroy Saint-Hilaire 31 août 1859 (écriture typique)	Lettre d'Isidore Geoffroy Saint-Hilaire 31 août 1859 (écriture atypique)	Lettre de Félix-Archimède Pouchet 11 novembre 1860
d	*dont*	*d'une*	*discord*
f	*fin*	*fraternelles*	*ferai*
j	*je suis*	*je*	*Je*
p	*peut*	*perte*	*pouvoir*
t	*tomber*	*toujours*	*telle*
x	*choix*	*mieux*	*yeux*
C	*C'est*	*Cette*	*Cosmos*
E	*En*	*Et*	*En*
P	*Paris*	*Ponts*	*Pouchet*
Q	*Quoique*	*qui*	*Quoi*

Plate 6. Comparison of the handwriting (Geoffroy Saint-Hilaire, Pouchet).

of scientific deontology. Prisoner of his own convictions, or prey to an overwhelming sense of resentment, Pouchet allowed himself to go so far as to falsify documents.

A letter of Isidore Geoffroy Saint-Hilaire, dated August 31, 1859, represents a significant piece of evidence in this regard. Its contents give one to understand that the Alhumbert prize of 1862 had been specifically created to honor the work of Pouchet. Let us examine this letter more closely (Plate 3). What strikes one immediately is that the handwriting does not correspond to that of Isidore Geoffroy Saint-Hilaire (Plate 4). Therefore, this letter cannot be an autograph. The handwriting is instead similar to that of certain official correspondence originating from the Rouen Museum of Natural History, such as the letter to the Minister of Education dated May 23, 1863 (BnF, naf 18111: 184). Thus, it may be hypothesized that the letter purported to be from Geoffroy Saint-Hilaire was actually written either by a secretary at the Museum in Rouen or by Félix-Archimède Pouchet himself. Let us compare the document with one known to have been written by Pouchet, such as the letter sent to Joly on November 11, 1860 (Plate 5). An examination of the handwriting shows that the letter of August 31, 1859, was not written in a smooth and regular manner; there are signs of hesitation, as well as variations in the form of certain letters. If one compares specific characters in this atypical autograph with the same ones in the letter written by Pouchet, it becomes clear that the two documents were penned by the same hand (Plate 6). Pouchet therefore was the author of the letter supposedly sent to him by the academician Geoffroy Saint-Hilaire on August 31, 1859.

Could this document be the legitimate copy of an original that had somehow been damaged or destroyed, or is it completely false? This point cannot be established by an analysis of the handwriting. However, the style of the letter provides some indications; one notes the use of certain turns of phrase that never appear in any of the autographs of Isidore Geoffroy Saint-Hilaire. Systematic evaluation of a sample of his correspondence (Rouen Museum of Natural History, FAP 3102) reveals the stylistic traits of the author. His manner is direct and not devoid of a certain finesse. He expresses himself with clarity and precision, and never indulges in long disquisitions. Certain passages in the letter of August 31, 1859, are therefore atypical of the pen of Geoffroy Saint-Hilaire, where one observes, in addition to tiresomely long passages not at all in conformity with his manner of expressing himself: 1) the use of superlatives and hyperboles that lend a grandiloquent tone to author's thoughts;[7] 2) a reference to the convictions of the man of science;[8] 3) an unwaveringly positivist tone that is foreign to the presumed author of the letter;[9] and 4) a paean to fallen heroes, the victims of injustice[10]—an image that Pouchet never ceased to promulgate on his own account after 1855, the year in which he failed to win the position of professor at the Paris Museum of Natural History.

I reproduce here some passages from this letter:[11]

My dear and most honored colleague,

[After listing, by way of apology, a list of impediments, Geoffroy Saint-Hilaire continues] This is why I have not yet advanced very far in the reading of your work. I have just finished this morning with the metaphysics. Yesterday I read the historical part. I believe that one could not have better presented, than by these two extremely learned and philosophical articles, the question that you treat; this is the most proper way to introduce the question in all its grandeur, and to sweep away all of the extraneous elements that have complicated it [. . .] I will resume your book this evening and tomorrow morning. I am eager to arrive at your experiments. I have not yet taken a position here. I believe what you tell me of the creative power that continues in time, very wise, very elevated and in the true sense, very religious. Spontaneous generation is, as you so well put it, rejected by the majority based on the authority of the expert. However, in science the authority of the expert is a weak argument [. . .] Let us come therefore to the facts, to the experiments; herein lie the *true experts*. I myself contributed in no small part to bringing forward the proposal for the prize that is to be offered by the Académie, the selection of which must have brought you satisfaction and flattered you greatly; because, before you submitted your scientific communications to the Académie, who would ever have thought, who would ever have dared to propose such a topic to them? It is M. Flourens who deserves the credit for having brought it forward; but I can do myself this justice [and say] that its adoption was due to me; because our three other colleagues, and as a consequence the majority, did not wish to give the impression that there were any *doubts*, and M. Flourens who does not much enjoy arguments, allowed his proposal to be laid aside. It is only that it was necessary to find a formula that did not in any way commit the Académie; and here as well he had the wisdom not to commit himself. This question, in fact, is of the most immense profundity, and the further one penetrates the more one is convinced [. . .] I would also like to tell you that I read your preface with a most sincere sentiment of sympathy; you mention some in order to thank them; you remain silent on others regarding whom you would have every right to complain: this moderation does you honor, and in your rising all the higher above such an injustice, which we have not forgotten, you demonstrate all the more clearly <u>how great it was and how unjustified it remains</u>. Adieu, I beg you to accept the expression of my most distinguished and devoted sentiments.

This exalted rhetoric, with its references to grandeur and to the fight against injustice, is entirely unlike the letters written by Isidore Geoffroy Saint-Hilaire (Rouen Museum of Natural History, FAP 3102). Stylistic analysis therefore establishes that Pouchet manipulated the contents of this particular document, in the process leaving clear traces of his intervention. All the same, the letter does contain some details pertaining to Geoffroy Saint-Hilaire that Pouchet could not have invented; the correspondent confesses himself to be weighed down by the task of correcting galley proofs, he was thinking of showing *Hétérogénie* to the son of Jean-Baptiste de Lamarck, who was in fact attending his lectures at the time, he expresses regret that Charles Bonnet was not cited, and so on. The conclusion which must be drawn therefore is that Pouchet did receive a letter from Isidore Geoffroy Saint-Hilaire around this time, but falsified the contents to his own advantage. Whatever the extent of his intervention may have been, this document is clearly a counterfeit. The incident serves to highlight with even greater force the question of professional and personal ethics in the conduct of science (Merton 1973) and how this was transgressed by Pouchet[12] (Raynaud 1999).

If one follows the inventory of asymmetries drawn up by Latour (1989), there was a net advantage on the side of Pasteur. The settlement of the controversy was decided by his scientific prestige, his powerful position at the Academy of Sciences, and the protection of politicians, not to mention the underhanded "maneuvers" that he supposedly engaged in to better his rival. A detailed examination, however, reveals that the status of the asymmetries outlined by Latour are much less clear; one finds evidence of Shinn's assertion that the relativists do not take into account many nuances (1994, 79). Moreover, certain asymmetries that are crucial from a sociological perspective have been passed over in silence. Pouchet transgressed the norms of the scientific ethos on more than one occasion because of his dogmatism, his penchant for carrying the debate into the unofficial forum of the press rather than limiting it to the constitutive forum, and even engaging in deception—all of which make Pouchet a close homologue to the authors of the most well-known contemporary scientific frauds such as Long (1979), Chu (1987), Gupta (1991), Herrmann (2000), Henninger (2007), and Hauser (2012).

One may now attempt to address the question: How is it possible that two studies in the sociology of the sciences arrive at such different conclusions? It should be noted first of all that Latour embraced Bloor's principal of symmetry (1983), but he adopted it for a purpose that was, to say the least, idiosyncratic: "The winners do not need to be *protected* by the historian, but the losers do, to whom one will grant [...] *a second chance* before the *tribunal of history*" (Latour 1989, 430, italics mine). One observes here a surprising reading of the principle of symmetry. According to Bloor, the principle signified that one must treat beliefs, whether they are true or false, in the same way. This was deducible from the principle of impartiality, which requires that one be "impartial with respect to truth and falsity, rationality and irrationality, success or failure" (1991, 7). And yet the very notion of "defending the vanquished" contradicts the principle of impartiality—above all when the loser resorts to scientific fraud. One might discern an analogy between this way of writing history and the penning of "hagiographies"—justly criticized by the advocates of SSK—the difference lying in a simple inversion, where one is no longer defending the "strong" but the "weak." The desire to redress the balance between the actors in the controversy therefore becomes artificial and self-refuting in nature: the sociologist cannot claim that he is reducing asymmetry when he is creating new ones. Even more, if the discovery of new asymmetries shows that in truth the "weak" were at the time more powerful than the "strong," what does the desire to "protect the vanquished" actually signify?

The same observation can be applied to the notion of the "tribunal of history." The hesitation among relativists between historical description and historical judgment is all the more surprising as this question has been amply discussed in sociology. Max Weber attaches great importance to the difference between the "empirical sciences" such as history and sociology and the "axiological sciences" such as aesthetics and law. The German sociologist writes that ethical evaluations "are problems in axiology, not in the methodology of the empirical disciplines" (1964, 382). Should sociology take the side of Pouchet or Pasteur? The verdict of Weber would probably have been contrary to that of Latour, based on the insurmountable difference between what "is the case" (*Seiende*) and what "ought to be the case" (*Seinsollende*) (Weber 1964, 428).

3. The Social Conditioning of Science

The sociologists of relativist obedience have endeavored to study the ideological and religious background to this controversy, but the diversity of their positions suffices to suggest that the debate regarding this aspect is far from being closed (Crosland 1992, 431). Latour asserted that the refutation of the thesis of spontaneous generation, which sprang up in the midst of the ongoing quarrel over transformism and creationism, had theological underpinnings in which Pasteur embodied the position of the spiritualist creationist (1989, 433). Vinck analyzes this opposition and its consequences in these terms: "Neither the theory of spontaneous generation, nor Darwinian theory [. . .] were politically acceptable. Therefore the scientific elite of the country was in need of arguments to reject these materialist theories" (1995, 114).

This interpretation hardly differs from that of Geison, apart from one crucial point (highlighted in italics by me): "Pouchet's *results* were cited to support materialism, evolutionism, and radical politics, while the opposite *results* of Pasteur were utilized to support spiritualism, the biblical explanation of creation, and conservative political policies" (1981, 371). One notes here a shift in interpretation, justly marked by the word *results*. Sociologists themselves may reflect, each in his own manner, the divide between materialists and spiritualists, radicals and conservatives, evolutionists and creationists, but while some believe that ideological factors conditioned the scientific debate (society → science) over spontaneous generation, others detect the recovery and confirmation of ideological concepts from the scientific results (science → society). It is therefore appropriate to return to an analysis of the social factors involved in this controversy, if only to *determine the meaning of the relationship* that linked science to its social context.

One can assign a second role to this analysis as well; while the portraits of the protagonists traced by Farley, Geison, Latour, Collins, Pinch, and Vinck are all quite consistent, other scholars have underlined the fact that the debate between Pasteur and Pouchet was being conducted against the backdrop of discussions over larger issues, and the relationship between spontaneous generation, evolution and the existence of God was itself a subject of controversy during the nineteenth century. As the historian of medicine Darmon writes: "Some saw in the ideas [of Pasteur] a justification of the existence of God. Others accused him of atheism [. . .]. The Abbé Moigno claimed that the theory of the pre-existence of germs was capable of converting non-believers and atheists" (1995, 148). The existence of such opinions does not invalidate the relativist interpretation, but it does motivate us to study *the degree of certitude* with which one can affirm that spontaneous generation was seen as an argument in favor of evolutionism rather than of the opposing thesis.

1. Pasteur and Pouchet

Relativist sociologists describe Pasteur as a spiritualist Catholic who, by refuting spontaneous generation, provided support for the biblical notion of creation.[13] If this were indeed the case, then one should find evidence of the opposite position in the texts of Pouchet. And yet, as Farley and Geison acknowledge explicitly, "It is manifestly absurd that one could have associated [Pouchet] with materialist, transformist and atheist forces" (1991, 103). In *Nouvelles expériences sur la génération spontanée*, Pouchet (1864, 199) declared himself to be opposed to Darwinism and a believer in fixism. It was he who in *Hétérogénie* (1859) launched an attack on materialism and shaped his thesis to conform to religious dogma:

> If the phenomenon [of spontaneous generation] exists, it is God who wished to employ it for his own ends [. . .]. The theories of the heterogenists, far from vexing the attributes of the Creator, merely augment His divine majesty. If, sometimes, in the silence of his laboratory, the man of science produces the evolution of some new being, his sense of pride will not allow him to delude himself; he knows that he is nothing more an intelligent journeyman who is realizing the conceptions of the sublime Master (1859, 97–98).

While Pouchet supported his thesis of heterogeny with proofs derived from his religious orthodoxy, Pasteur would never have considered resorting to such devices in his argumentation (Roll-Hansen 1979, 276). As Geison himself recognizes: "At the center of Pasteur's public views on religion and philosophy lay his insistence on an absolute separation between matters of science and matters of faith or sentiment" (1981, 354). And it was Pasteur who declares in the lecture on spontaneous generation that he delivered in 1864: "Here it is not religion, or philosophy, or atheism, or materialism, or spiritualism that hold [. . .]. Here it is a question of fact" (1922, 345).

The examination of the ties between Pouchet's personal convictions and the thesis of heterogeny can be taken even further. It is not accurate to claim that, "Just how [Pouchet] arrived at making this problem his principal focus of interest remains obscure" (Farley and Geison 1991, 98). In fact, the articles that he writes on the generation of animals before 1848 were opposed to the theory of heterogeny. In a note published in the *Comptes rendus de l'Académie des Sciences* in 1848, Pouchet claimed that *Infusoria* were animals that engaged in sexual reproduction. Pouchet therefore changed his position between

1848 and 1859, the period during which he was working on *Histoire des sciences naturelles au Moyen Âge* (1853). Could this book have contributed to his change in attitude towards spontaneous generation? There is every reason to believe so.

The medieval notions presented in *Histoire des sciences naturelles au Moyen Âge* shed light on the beginnings of Pouchet's interest in heterogeny. The naturalist from Rouen cites various works that treat the subject of spontaneous generation, such as Aristotle's *De generatione animalium*, Albertus Magnus's *De parvis animalibus*, and Agricola's *De animantibus subterraneis*. Oddly enough, however, he does not elaborate on the question himself.[14] Was Pouchet reserving a more thorough examination for a future study?

Histoire des sciences naturelles au Moyen Âge was an apologia for Albertus Magnus,[15] whom the author described as "the finest mind of the entire period," and sheds light on the origins of the quasi-religious manner in which Pouchet would later address the question of spontaneous generation:

> To begin [Albertus Magnus] seized upon the study of nature in order to reinforce the science of God or theology [. . .] It is in this way that, for the first time, a savant, embracing the universality of both the human and the sacred sciences, elevated himself to the sublime by tracing *the relationships between man and God!* [. . .] Presented in this way, the natural sciences could appear in their fundamental character: their physical utility and their theological utility (1853, 214–215, 320).

Pouchet's book would be criticized by Daremberg (*Journal des Débats*, January 16, 1854). Despite such attacks, this rehabilitation of medieval thought would profoundly influence the work of the naturalist. After Pouchet's death an inventory of his books was compiled and it testifies to his eclecticism. His library contained thirty-seven works on spirituality (from symbolism and alchemy to demonology, etc.), whereas the whole of "medicine, anatomy, and physiology" was covered by fifty-four works, a surprisingly modest number if one considers that Pouchet was a doctor of medicine. Given these conditions, one would be inclined to conclude that Pouchet became interested in the question of spontaneous generation following his discovery of the medieval theory of heterogeny, referring to the model of Albertus Magnus which, as we will see below, was largely articulated within the framework of Christian theology.

The first difficulty that one encounters in assigning evolutionist or materialist values to the thesis of spontaneous generation is that Pouchet—contrary to what has been claimed by the proponents of relativism—relied explicitly on a creationist vision in his *Hétérogénie*. The controversy over spontaneous generation therefore cannot be viewed as the opposition between a conservative, spiritualist partisan of the biblical story of creation (Pasteur) and a radical materialist and partisan of evolutionism (Pouchet). Such a picture would prevent us from understanding the true foundations of the controversy over spontaneous generation.

2. The Family of the Heterogenists

The difficulties of associating the thesis of spontaneous generation with certain political or religious ideologies also emerges from an examination of the intellectuals who participated in the debate beginning in the seventeenth century. Neither the partisans of heterogeny nor the partisans of homogeny formed homogeneous groups.

Some men of science rejected the theory of spontaneous generation *a priori* on religious grounds. Jan Swammerdam for example was firmly opposed, because it would have required him to acknowledge the role of chance and in this way would have challenged the creationist plan in his *Historia insectorum* (1669). Pierre Lyonet, the author of *Théologie des insectes ou démonstration des perfections de Dieu* (1742), was a convinced Calvinist. Spallanzani, to whom we owe the first demonstration that organic infusions brought to the boiling point in a hermetically sealed ampoule remain *ad infinitum* devoid of microbial life (1765), was a Catholic priest.[16]

However, there were just as many savants with a religious vocation who defended heterogeny. The Dutch artist Johannes Goedaert, who published a series of insect studies in *Metamorphosis naturalis* (1658), belonged to the Reformed Church. And it was an Italian Jesuit, Filippo Buonanni, who countered the experiments of Francesco Redi with his own *Observationes circi viventia* (1691). The celebrated biologist John Turberville Needham was ordained a priest of the secular clergy twelve years before publishing *Nouvelles observations microscopiques* (1750). According to Horwitz Westbrook (1981, 9), his scientific studies were motivated by the "desire to defend religion at a time when biological questions had theological and philosophical implications." The naturalist Jean-Claude de Lamétherie laid out arguments in favor of spontaneous generation in *Principes de la philosophie naturelle*

(1787) without however renouncing creationism. As for Erasmus Darwin, who adopted a position favorable to spontaneous generation in *Zoonomia* (1796), he would defend a religious conception of the universe in *The Temple of Nature* (1803). Finally, what can be said about Carl Wilhelm von Naegeli (1817–91)? This eminent botanist, who conducted important studies on cell division and pollination, would remain attached to the thesis of spontaneous generation for his entire life even though he was a determined opponent of Darwin's ideas on natural selection.

This list of anomalies, which could be considerably extended,[17] shows that the association between heterogeny and evolutionism imagined by certain sociologists of the sciences to account for the controversy between Pasteur and Pouchet is very fragile indeed. If intellectuals from Goedaert to Buonanni and Needham, from Lamétherie to Darwin, did not subscribe to such an association, it was because it did not constitute an imperative. In any event, the designation of Pasteur as a creationist reveals itself at best to be based on a "personal correlation" with no connection to the theory of the social determination of scientific knowledge, and at worst a grave error of assessment which perceives the influence of social factors where in fact there was none.

One of the problems therefore in associating evolutionism and spontaneous generation rests therefore on the diversity of opinions held by naturalists and microscopists. There were evolutionist heterogenists just as there were creationist heterogenists, evolutionist homogenists just as there were creationist homogenists, and hence there was a range of opinions from which one cannot reasonably draw any conclusions. As a last resort, one might say that it is the venerable antiquity of the thesis of spontaneous generation in the Christian world that explains the existence of these concurrent interpretations.

3. Theology and Spontaneous Generation

To understand how a deist and creationist interpretation could have been deduced from or reconciled with the thesis of spontaneous generation, it is necessary to go back to a time when the alternative theory of evolution had not yet been conceived.

In studying the long history of the belief in heterogeny, the first thing one finds is the progressive reduction of the thesis to ever more limited categories of living beings. Aristotle spoke of spontaneous generation with reference to the insects, mollusks, and certain fishes such as the eel

(*De generatione animalium*, III, 11). As the study of natural history was revived in the Middle Ages, Petrus Bonus Lombardus pointed out that Aristotle's interpretation did not hold, at least in the case of frogs and insects (*Theatrum chemicum*, V). In the seventeenth century the thesis was reduced to insects and then, beginning in the eighteenth century, to microorganisms. Indeed it would not be false to say that the infinitely small and barely detectable have always provided a loophole for the thesis of spontaneous generation, in parallel with the development of the instruments required to visualize them. But despite these changes in scale, one finds the same hypotheses turning up repeatedly—that certain living things possess the capacity to form spontaneously from an environment rich in organic material, and under the effect of factors such as heat or light.

An examination of the ties between heterogeny and the theories on the origin of species requires us first of all to recognize that evolutionism was not the only revolution that the thesis of spontaneous generation was exposed to. The doctrines of the classical authorities Aristotle and Theophrastus were based on the notion of the fixity of species (which will remain the same for all eternity), and therefore the first great revolution was that of creationism (which posits that all species have remained the same ever since they were created by God). In this period, therefore, various attempts were made to draw the thesis of spontaneous generation closer to the scenario of creation.

The first religious justification of heterogeny that Pouchet offered was based on a somewhat forced interpretation of the biblical account of Samson finding a nest of bees in the carcass of a lion. Pouchet writes: "In *Judges*, the inspired author imagined that a swarm of bees was born from the decaying entrails of a young lion and Samson robbed them of their honey for a feast (Pouchet 1859, 10). A comparison of this assertion with the biblical text, whether in its Hebrew, Greek, or Latin version, shows that Pouchet had engaged in a broad "over-interpretation" of the passage in order to harmonize it with his own vision of heterogeny.[18]

He found a second justification in the texts of the Fathers of the Church, and in particular the writings of Saint Augustine which contain, Pouchet (1859) says, "manifest proof that the claims of the heterogenists have never contravened orthodoxy" (103). He cites the following passage from *De genesi ad litteram*: "It has been observed, you see, that some animals are born of water or of earth without having any distinction of sex; and thus their seed is not in them, but in the elements from which they arise by spontaneous generation."[19] According to Pouchet, all that remained was to determine whether the creation of these germs was contemporary with their appearance in the world of the living, or whether it predated them. Saint Augustine writes: "There is also a real question about the creation of the minutest animals, whether they were created in the original establishment of things, or from the putrefaction consequent upon material things being perishable. Most of them, you see, are either bred from the sores of living bodies, or from garbage and effluents, or from the rotting of corpses; some also from rotten wood and grass, some from rotten fruit; and we cannot say that there are any of them of which God is not the creator [. . .] And it can indeed be said that the tiniest things which spring from the waters of the earth were created then."[20] Thus, the smallest animals were created at the very origin of all things. To better understand this argument, it is necessary to recall that, according to Saint Augustine, God undertook to simultaneously create all the "seeds" of life, some of which were then further developed. This schema made it possible to conceive a process of creation by stages without compromising the notion of a divine origin for all living things. Within this framework the appearance of a new species did not present an obstacle to the biblical scenario of creation. The objective of Pouchet's "seminal reasons" was to permit this reconciliation.

The justifications outlined here show why spontaneous generation could still, in the second half of the nineteenth century, recruit many supporters among theologians.

4. Conclusions

A detailed examination of the controversy between Pasteur and Pouchet allows us to draw conclusions of some importance with regard to the methodology of SSK.

In their analysis of the debate the principle of accumulated asymmetries was applied, but some of the asymmetries identified must be considered liable to revision; for example: 1) Pouchet maintained ties of friendship with several members of the Academy of Sciences, some of whom acted as judges of his work; 2) He had known them for twenty years, long before Pasteur became a member of the same academy; 3) Pouchet mobilized prominent figures outside the scientific community to intervene in his disputes with Milne-Edwards and Pasteur, seemingly without success; 4) He resorted to illicit means in order to gain an advantage over his rival.

Other asymmetries have been deliberately overlooked: 1′) At the time of the controversy Pasteur, not being attached to a research institution, did not have a secure scientific status and position; 2′) Pouchet's age does much to explain his sense of injury and resentment; 3′) Pouchet had a more prolific output than Pasteur, but his texts were not limited to reporting scientific results; 4′) Pouchet was not a good experimentalist; 5′) His passionate commitment to his ideas differed from Pasteur's skepticism; 6′) Pouchet's regrettable lapses in terms of scientific deontology contrast with the probity of his opponent.

This new inventory of asymmetries completely reverses the SSK vision of the balance of power between the two sides in the controversy. The difference is all the more striking if one considers that certain analyses have tended to adopt the approach of the "defense of the loser," relying on the biography of Pouchet by Pennetier (1907) as a source. Pennetier was a disciple of the Rouenese naturalist and his account is based on the letter of Isidore Geoffroy Saint-Hilaire, which—as we have shown—was a forgery.

Finally, there is no evidence that the experts called in to adjudicate on the question of spontaneous generation succumbed to unfair pressure in making their final decision. If Pouchet emerged the loser in this controversy it was first and foremost because: 1) he withdrew from the competition for the Prix Alhumbert in 1862; 2) in 1864 he refused to present himself before the commission that had been formed at his own request [39]; and 3) his colleagues at the academy seemed to have tired of his unethical behavior. In attempting to settle a question of science by nonscientific means—campaigns of denigration, appeals to the Ministry and the press (what Pasteur referred to with irritation as "the justice of the daily papers")—Pouchet strayed irredeemably beyond the boundaries of the norms of "disinterestedness" and "organized skepticism" (Merton 1973, 275, 277).

The relativist thesis that there is no demarcation between science and nonscience leads to its own transgression when SSK introduces the factors of politics and religion into the debate over spontaneous generation. The presumed gap between heterogenists (radical evolutionists) and homogenists (conservative creationists) is anything but obvious: Pouchet himself advanced proofs that his thesis was in conformity with received Christian dogma; the partisans of spontaneous generation formed a diverse family that contained as many atheists as Christians; and biblical sources were drawn upon for arguments that would buttress the thesis of spontaneous generation.

These critiques certainly do not undermine the principle of accumulated asymmetries. But it is excessive to believe that the results of this type of study commit one to adhere to relativist principles. The possibility that Pasteur's Catholic faith and political conservatism might have influenced him in the formulation of his thesis of homogeny has no real significance. There are two main reasons for this.

1. As pointed out above, all of the facts reported by Farley and Geison (1991, 126) in support of their thesis that political influence was a determining factor in the resolution of the controversy have been dated with deliberate vagueness to "the 1860s." Yet Pasteur received his first invitation to the royal Chateau de Compiègne in 1865, his studies on wine and on the silkworm (financed by Napoléon III) were conducted in 1865 and 1870, and he was nominated a *commandeur* of the Légion d'Honneur in 1868. Therefore, none of these events, all of which date to after 1864, could have exerted any influence on the settlement of the controversy. The anachronism is even more striking in the following account: "The oppositions [to Pasteur] were crushed by political maneuvering, by ridicule, and by Pasteur drawing farmers, brewers, and doctors to his cause" (Collins and Pinch 1993, 80). This was not true of *farmers*, because Pasteur studied chicken cholera in 1878, sheep anthrax in 1881, bovine pleuropneumonia in 1882, and swine erysipelas in 1883. This was also not true of *brewers*, because Pasteur registered his two patents, one on an improved process for manufacturing beer and the other on the production of brewer's yeast free from organic germs in 1871 and 1873, respectively. Finally, this was not true of *doctors*, because Pasteur's work was hailed by physicians only after he developed his antirabies vaccine in 1885. Since these events all date to the years 1871–85, none of the interest groups mentioned could have played a role in the closure of the controversy, which unfolded between 1859 and 1864.

2. As soon as one examines more closely the family of believers in homogeny, one notes such a diversity of positions that the links between homogeny, politics and religion turn into personal correlations that have no real contribution to make to our understanding of the controversy. Similar critiques may be applied to the study of the supposed relationship between eugenics and statistics (Norton 1978). The larger the population studied, the more what may appear to be a *determining* correlation on the scale of the individual becomes insignificant. For example, not all eugenicists are statisticians and not all statisticians are eugenicists (see Chapter 6); in the same way, Donald MacKenzie (1991) recognizes that "not all the members of the Biometric School

were convinced eugenicists" (243). This fact raises the question of the actual explicative power of the social factors that are brought into play in the "demonstrations" presented by SSK. Sociologists often confuse "mere coincidences" with "regular features", thus weakening their analyses. This is why a rereading of the debate between Pasteur and Pouchet prompts us to revise the very principles that govern the sociological analysis of scientific controversies:

—The scrutiny of the asymmetries in a debate must distinguish between the various arguments as a function of the forums in which they appear. Debates that take place in the contingent forum are often the consequence, and not the cause, of those conducted in the constitutive forum.

—The survey of asymmetries must be as exhaustive as possible, because the balance that is capable of explaining the success of one side over the other will depend very closely on their selection. The asymmetries must always be sought for in the community of researchers in such a way as not to mistakenly identify as *typical* certain correlations that are actually *personal or fortuitous*.

Notes

1. It may be noted that initially the commission was to have included Isidore Geoffroy Saint-Hilaire and Augustin Serres. However, Geoffroy Saint-Hilaire passed away in the interval and Serres withdrew; they were replaced by Claude Bernard and Victor Coste.

2. For instance, Pasteur's salutations to Balard betray no hint of close personal or professional ties: "Will you accept, Sir, the expressions of respectful devotion with which I have the honor to be your very humble and very devoted servant." These reserved and formal greetings contrast with the formulae that Pasteur employed in his letters to Dumas ("Your very devoted and grateful servant") and to Raulin and Bertin ("I am most sincerely yours").

3. It is known, for example, that not all of the members were in the habit of frequenting one another outside of the academy. Thus, when a certain Julia asked Geoffroy Saint-Hilaire to intercede with Flourens on her behalf, the scientist responds: "I am not in a position to introduce you, not having regular contacts with him" (BnF, naf 16320: 159).

4. Pouchet assiduously cites in *Hétérogénie* all of the more influential members of the Academy of Sciences, many of them more than once, as a rapid count limited to the authors who appear in the footnotes shows. In decreasing order of frequency we find the names: Geoffroy Saint-Hilaire (11), Blainville (8), Quatrefages (8), Dumas (6), Milne-Edwards (6), Brongniart (5), Bernard (5), Rayer (3), Serres (2), Flourens (2), Chevreul (2), Lacaze (2).

5. Ian Hacking proposed the notion of "styles of reasoning" as a substitute for the term *Denkstil* or "style of thinking" (*Denkstil*) which had been adopted

early on by Fleck (1935), and then by Mannheim (1956) and Crombie (1994). Hacking (1992) explains the reason for his modification as follows: "I choose 'reasoning' over 'thinking' because I am more concerned with what is said than what is thought" (138).

6. The adherents to SSK often evoke Pouchet's talents as an experimentalist, but never in any detail and one can understand why. Not only did Pouchet claim to have found living microorganisms in an infusion that had been heated to 250°C, he (1859) cites "unimpeachable reports of rainfalls of frogs" and "avalanches of stickleback fishes" as proof of his theory of heterogeny (448). He (1859) also believed that he had demonstrated experimentally that red light promotes the formation of animal protoorganisms and green light promotes the formation of vegetable proto-organisms (197). Pouchet did not make any distinction between hypotheses and experimentally demonstrated facts.

7. Pouchet's texts were liberally sprinkled with superlatives such as "articles si savants et si philosophiques" and "très sage, très élevé et très religieux" [7]. In his letters to his colleague Joly one finds such extravagant adjectives as: "in such an extraordinary manner; such peculiar facts; such magnificent support" (BnF, naf 18111: 8, 13, 18, 21). The same formulae appear in *Hétérogénie*: "The ever so haughty reprobation; a more magnanimous and more disciplined army" (1859, 5).

8. Typical of Pouchet is this lofty passage praising the tenacity of the savant, who "lives for science, will sacrifice everything to her; science has her own paths, and she has the right to follow them without worrying about anything other than the search for the truth." He writes to Joly: "Courage, therefore, and march forward, inasmuch as we have the conviction of being in the right; Courage, my noble friend, you must not allow yourself to be beaten; I personally will never give up the battle" (BnF, naf 18111: 21, 50, 77 & 123).

9. The positivist tones ("the experiments; herein lie *true experts*") formed an integral part of Pouchet's rhetoric. This is attested to in his correspondence with Pasteur: "Chemistry, however powerful it may be, will never annihilate such facts; when it is a matter of subtle observations" (BnF, naf 18106: 47, 49). In the preface to *Hétérogénie*, Pouchet (1859) informs the reader that his book is composed of two parts: "the experimental part, which is the most fundamental, and the theoretical part, which only forms an accessory fragment" (vii).

10. His tributes to the victims of injustice ("The justice that you render . . .; so unjustly treated . . .; the beautiful justice that you render . . .; rising above such injustice . . .") spread to his correspondence with Joly: "You will have justice," he writes (BnF, naf 18111: 47). Pouchet also complained, in a letter written in 1864, of "the distinctions with which the Minister honors so many of those academic nullities; piling all of its favors on the little corner of Paris alone." In his *Histoire des sciences naturelles*, Pouchet (1853) had already written in the same fervent tones, "We will defend the Middle Ages from the unjust reproaches of which it has so often been the object" (7).

11. This letter [7] is conserved in the Rouen Museum of Natural History, FAP 3102. The most dubious passages from the point of view of authenticity are underlined (the italics instead are Pouchet's).

12. In the same vein, it may be noted that in the exemplar of *Nouvelles expéri-ences* presented by Pouchet to a mysterious brother in the Freemasons, "Laki, Knight of the Persian embassy" (B.m. Rouen, ms. 3690), Pasteur's name has been crossed out several times and replaced with the words "some chemist" or "this chemist." Elsewhere Pouchet (1864) has cut the name of his adversary out of the page with a razor blade! (xiv, on the meeting with Milne-Edwards; 13–14, upper part of the page; 87–88, lower part of the page). As far as Pouchet was concerned, his ambition to continue playing the match against Pasteur was as strong as ever in 1864.

13. According to Mendelsohn (1987, 110) and Farley and Geison (1991, 119), all of whom relied on the short biography by Georges Pennetier (1907) as their source, Pasteur's support on the commission of the Academy of Sciences came from members who were convinced Catholics and adherents to the theory of fixism. Yet, as Ernan McMullin observes (1987, 87), this hypothesis is difficult to reconcile with certain facts, such as that Claude Bernard (who was not a Catholic) was one of the most vociferous critics of Pouchet's theory of heterogeny, and the then secretary of the Academy of Sciences, Armand de Quatrefages, maintained close relations with Darwin, which were cordial enough that Darwin once brought his son to meet him (BnF, naf 11824: 72).

14. Gálvez (1988, 362) plausibly argues that Pouchet may have drawn his notions on heterogeny from his reading of the works of Albertus Magnus. Regret-tably, this cannot be established due to Pouchet's silence on the point.

15. For that matter Pouchet (1853) quotes repeatedly from *Le Génie du chris-tianisme* whose author, Chateaubriand (1802), was an ardent promoter of medievalism. From this, and perhaps under the influence of Blainville, Pouchet's thought began to reflect aspects of the Christian spiritualist move-ment and its vision of the Middle Ages as an intensely spiritual and heroic period.

16. Studies using the microscope would lead in the middle of the eighteenth century to a celebrated controversy between Spallanzani and Needham. Needham contested Spallanzani's experiments, because the air hermetically sealed in his ampoules and subjected to heating could have been altered to the point that spontaneous generation became impossible (Geison 1981, 367). This objection acquired renewed force when Gay-Lussac demonstrated the role of oxygen in the processes of fermentation and putrefaction in 1810. And in 1837 Theodor Schwann would show that infusions remained unaltered after contact with air, if this air was first calcinated.

17. Analogous anomalies can be detected among the adversaries of spontaneous generation. The American naturalist Jeffries Wyman (1814–74) demon-strated that life could not develop in infusions that had been brought to the boiling point, but in private admitted that he believed in Darwinism (Doesch, 1962). Similarly the German microscopist Ehrenberg (1795–1876) declared his opposition to the theory of spontaneous generation even though he was a supporter of the evolution of species.

18. "Samson tore the lion as he would have torn a kid, and he had nothing in his hand [. . .] And he went down, and talked with the woman; and she pleased Samson well. And after a time he returned to take her, and he turned aside to see the carcase of the lion: and, behold, there was a swarm of bees and

honey in the carcase of the lion" (Judg. 14:8). The Hebrew, Septuagint and Vulgate versions of the Old Testament provide such concordant readings that the thesis of a biblical case of spontaneous generation can be rejected. This was a case of pure invention on the part of Pouchet. The three biblical versions are unanimous in their use of the expression "there was . . . in," with no suggestion of a sudden and spontaneous apparition of bees. The fact that there was honey proves that the colony had been living in the carcass for a long period of time.

19. "Observatum est enim quaedam ita nasci ex aquis vel terra ut sexus eis nullus sit et ideo semen eorum non sit in eis sed in elementis ex quibus oriuntur" (*De genesi ad litteram*, III, XII, 19, ed. Rotelle 2002, 227).

20. "Nonnulla etiam de quibusdam minutissimis animalibus quaestio est utrum in primis rerum conditionibus creata sint an ex consequentibus rerum mortalium corruptionibus? Nam pleraque eorum aut de vivorum corporum vitiis vel purgamentis vel exhalationibus aut cadaverum tabe gignuntur quaedam etiam de corruptione lignorum et herbarum quaedam de corruptionibus fructuum: quorum omnium non possumus recte dicere deum non esse creatorem [. . .] Et potest quidem dici ea minutissima quae ex aquis vel terris oriuntur tunc creata" *(De genesi ad litteram,* III, XIV, 22–23, ed. Rotelle 2002, 229).

3

The Vitalism–Organicism Controversy between Paris and Montpellier: An Essay on the Social Determination of Knowledge

The debate between relativists and rationalists is not limited to the general principles of the sociology of scientific knowledge. It sometimes involves specific points in the relativists' theoretical program, such as the principle of causality (Bloor 1991, 7). In an analysis of the scientific content associated with a controversy, this principle signifies that the sociologist must look for a causal relationship between a state of the society and a given state of the sciences (which might be, for example, the influence of a political ideology or economic interests on a scientific theory). It is significant that the questions posed by the rationalists often touch upon the sense that one can attribute to the idea of a "causal determination of science," and that the variations to be seen in the position of relativists from Bloor to Collins reflect a softening of this relationship of causality. Are we dealing with determination, conditioning or mere homology? Our understanding of the nature of this relationship could gain much from a study of specific cases. We will examine here the controversy between the doctrines of vitalism and organicism which in the middle of the nineteenth century opposed the schools of medicine of Montpellier and Paris.

This controversy could be of interest to the sociology of the sciences because, on the local level, the history of the faculty of medicine in the city of Montpellier shows that the dispute rapidly passed beyond the realm of medical knowledge, and assumed the outlines of a polemic that did not exclude metaphysical and political arguments, professional interests, and personal jealousies. Such a rich stock of extrascientific elements could reveal whether external factors played a role in the settlement of the controversy and to what degree they assured the success of one theory over the other. It could in addition show whether the scientific positions of the participants underwent any shifts and if the truth in this case was indeed negotiated, as the relativists claim.

1. A Chronicle of the Controversy

French medicine in the nineteenth century was shaken by a violent controversy that opposed two doctrines: *organicism*, which was supported by doctors in Paris, most notably Jean-Nicolas Corvisart (1755–1821), François Broussais (1772–1838), René Laënnec (1781–1826), François Magendie (1783–1855) and finally Claude Bernard (1813–78) and *vitalism*, which was defended by the disciples of Paul Joseph Barthez (1734–1806) in Montpellier, among them Charles-Louis Dumas (1765–1813), Jacques Bérard (1789–1829), Alexis Alquié (1812–64), and above all Jacques Lordat (1773–1870). These diametrically opposed positions would garner support that was often deeply entrenched, although it must be recognized that certain savants who were not closely associated with either school, such as Xavier Bichat (1771–1802) and Antoine-Louis Dugès (1797–1838), would attempt to reconcile the two doctrines.

By virtue of its multifactorial nature, it is no easy matter to pin down the exact dates of this controversy. It extended more or less from the death of Barthez in 1806 to the publication of *Introduction à l'étude de la médecine expérimentale* by Claude Bernard (1865). The most heated period began with the appearance of Broussais' *Examen des doctrines médicales* in 1821 and ran until the publication of Lordat's *Accord de la doctrine anthropologique de Montpellier avec ce que demandent les lois, la morale publique et les enseignements religieux prescrits par l'État* in 1852.[1] Between these two dates a spike in activity can be detected when the controversy became inextricably mixed with the polemic that sprang up between Lordat and Louis Peisse in 1840–1843.

The controversy between Montpellier and Paris opposed the *vitalism* of the disciples of Barthez,[2] who believed in the principles of a holistic approach to the human body, which they viewed as an interconnected system of sympathies and synergies between organs, etc. (known as the *doctrine médicale de l'École de Montpellier*), and the medical theory espoused by their rivals in Paris, which the Montpellierians referred to as *organicism*, *sensualism* or more broadly *materialism* (Lordat 1840,

399). Lordat (1840) uses these terms to criticize from every possible angle the "materialist synthesis" of Pierre-Jean Georges Cabanis, the "naturalist medicine" of Philippe Pinel, the "extravagantly anti-medical ideas" of Broussais, and the "infantile hypotheses" of Bichat (412).

When studying a disagreement between two individuals (such as Pasteur and Pouchet) it is relatively easy to trace the complete history of the controversy. In contrast, it is impossible to follow all of the threads of a dispute between two groups of researchers who may have been primed by long-standing professional rivalries.[3] Nevertheless, a sampling of the polemical exchanges in this case can be provided.

Before 1820 the Parisians did little more than cast doubt on the vitalist doctrine. They avoided entering into an open polemic, even though this did not prevent Bérard (1819) from perceiving a threat looming over his school, as is attested to in the preface that opens *Doctrine médicale de l'école de Montpellier* (9).[4] Nonetheless, the criticisms from the capital remained veiled. In his *Propositions sur la doctrine d'Hippocrate* published in 1804, Laënnec simply warns against the systematic spirit in medicine (perhaps prompted by his uncle, Guillaume Laënnec, a physician who had earned his degree at the University of Montpellier):

> Systematic ideas are the most variable that there are in medicine; every school, every period has its own; and in general one repudiates an author in proportion to the distance that separates his theory from one's own. From this perspective, among all the authors Hippocrates is the one who must displease the least. Nowhere did he expound in a continuous manner his systemic ideas. . . . Sometimes he even seemed to doubt that medicine could ever have a constant method (6).

However, the tone in Paris changes in 1821. In *Examen des doctrines médicales* Broussais (1829) openly attacked the doctrine of Montpellier, deploring the pernicious influence of Barthez:

> Barthez, who very quickly seized for himself the scepter of medicine, has just forced pathology one step backward [. . .] Barthez, *homme de cabinet*, erudite, in possession almost of omniscience, founded it [medicine] on his readings rather than on his observations, separating it from the organs and carrying it off into the clouds (384–385).

Broussais went on to point his finger at "the oracle of Montpellier" (390), the "triviality" of a "useless principle" (394), the "vicious circle" within which Barthez had trapped himself (399), his "[therapeutic] indications, almost all of them bad" (406), and so on.

In 1825, in the preface to the second edition of his *Précis élémentaire de physiologie*, Magendie declared that the texts of the Montpellierians contain "certain unintelligible words, such as the Force or the vital Principle" (viii).

The tone heightened even further in 1830 when the *Gazette médicale de Paris* published a *feuilleton* (a supplement on a topical issue, attached to a newspaper) consisting of four letters by an anonymous author that criticized the medical school of Montpellier (161–163, 229–231, 335–339, 403–405). The author was often bitingly sarcastic, as the following passage shows:

> Before anything I will say a word to you about the school *en masse*. It is not without reason that I qualify it as a mass rather than a body. A body, as you know, signifies a regular ensemble of parts in perfect harmony. A mass, on the contrary, is a confused aggregate of material with no unity. Such precisely is the school in Montpellier, whether in its internal policy or its medical doctrines. In all of these regards it is in a state of veritable anarchy [. . .] There one discusses and one disputes; one shouts, one rants as if in a popular assembly. From insults one passes to threats [. . .] One would never run out of material if one had to complete the tableau of the discord in this school. [. . .] You would laugh at the singularity of the principles that some of these Messieurs wish to inculcate in their students [. . .] Anatomy, pathological anatomy, the accompanying sciences are generally far too neglected. The greater part of the professors, convinced that the organization and the agents of the still life only play a secondary role in the functions of the living body, have embraced this truth in order to believe themselves justified in not granting these sciences anything but very superficial attention. [. . .] Pathological anatomy is treated with no less disdain. [. . .] Their ignorance and contempt for the concrete notions of medicine push them to an excess of abstraction. [. . .] Having thus repulsed the fixed point which they should never have abandoned, that is to say the living organization, physiology and pathology have been lost in vagueness. From this [stems] the obscurantism of which these professors are fairly accused (161–163).

Informed of these attacks, Jacques Delpech (1772–1832), professor of surgery at the faculty in Montpellier, responds in the *Mémorial des Hôpitaux du Midi* (1830, 318–319), rising up against the "indecent diatribe" of the "prudently anonymous" correspondent.

Delpech was murdered by a patient in 1832 and applicants were invited to participate in a competition to choose his successor in January-February 1834. The seven candidates who applied for the post were required to present themselves before a jury consisting of the president Dugès; six members Lordat, Alire Raffeneau Delile, Fulcran Caizergues, Eugène Delmas, Fontaines, and Paul Pagès, and two alternates Antoine Duportal and Albert Saisset. Six of the seven candidates however signed an open letter denouncing the "intrigues and favoritism" that tainted the competition and demanded that it be suspended. Michel Serre (1799–1840) was the only candidate to appear before the committee and he was awarded the professorship due to the absence of other candidates. The *Gazette médicale de Paris* related the improper handling of this *concours* in 1834, allowing both sides to give their account of the proceedings, but making it clear—and this is what is of interest to us here—that the jury in Montpellier was not impartial.

> What becomes of the guarantee in this mode of nomination and its claimed equity and openness, if it can be thus vitiated in its very constitution by protestations and shocking scandals of this type? Above all, what does this classification of the candidates by examination signify when, from the very beginning of the competition, every chance and every hope for the [other] contenders can be removed? (118).

In 1840, after two decades of such exchanges, we pass to the acute period of the dispute, which was inspired by a single line from *Fragments de philosophie*, a work by William Hamilton that had been translated into French by the health columnist Louis Peisse (1803–1880).[5] In his preface Peisse (1840) deplores the lack of logic and methodology in French universities, illustrating his point with a biting note on the school of medicine in Montpellier (with which he was acquainted, having completed a part of his studies there).

> This reproach of sensualist and materialist tendencies is directed in general at the school in Paris [. . .], but one must make a more special exception for the school of Montpellier, which has always professed the opposing principles, and which is so attached to them that metaphysics has often made them forget the practice of medicine (cxxvi).

There followed a polemic between Lordat and Peisse which took the form of an exchange of *feuilletons*. One can follow the episodes as they unfolded in the *Gazette médicale de Paris* and the *Journal de la Société de Médecine-Pratique de Montpellier* between the years 1840–1843.

In his first reply, Lordat (1840) defended his medical school against the accusation that it had lost its way in metaphysical speculation;

everything in Montpellier, he declared, was based on empirical observation (411). He went on to explain that the school's approach was not
spiritualist, this being a label that others had applied to it, rivals who
were incapable of conceiving that there might be a fair middle ground
between materialism and spiritualism (401).

Peisse responded with a series of newspaper supplements, each
devoted to one of the criticisms that he had raised against the school
of medicine in Montpellier, namely: its intellectual immobility (1841,
113–119), its penchant for metaphysics (1843, 37–42), its lethargy
(1843, 53–59), and its dogmatic spirit (1843: 85–93). In February 1841,
Peisse attacked first of all Montpellier's immobility:

> Montpellier, being separated geographically from the large centers
> of scientific activity that have formed over the last half century, has
> found herself gradually in a sort of intellectual isolation. Rather than
> taking part in the important developments that are going on around
> her, or at least following then, she is satisfied to observe from afar and
> from on high; she considers them as nothing more than [...] a pass-
> ing storm in the midst of which her Doctrine, like the boat of Saint
> Peter *that must not perish*, can be buffeted but never submerged [...]
> Believing that medicine is finished, and saying that she is to be found
> in Hippocrates, is not a good way to make it advance. From this stems
> the mistrust of Montpellier, if not yours, Sir, of innovations (Peisse,
> 1841: 113–119).

In the January 1843 issue of the *Gazette médicale*, Peisse attacked
the metaphysical propensities of the school, quoting mercilessly from
his adversary's *Doctrine médicale de Montpellier* (1819). As Bérard
(1819) writes, "From a long time ago, [to philosophize] has been for
us a sacred usage, an inviolable custom [. . .] But I have had occasion
to fear that more than one of our students has responded with much
more confidence on certain points of dogma, in the manner of the
philosopher, than on such and such a medicinal formula or on such
and such a minute point of anatomy" (15–16). Peisse (1843) concludes: "If Bérard was so mistaken with regard to the habits and the
spirit of the school where he himself was nurtured and developed, to
the exemplification of which he consecrated his entire life, and if the
portrait which he has drawn of it is not a good likeness, I truly do not
know to whom we must address ourselves to have exact information
on this point" (39–40).

In a following piece published in 1843, Peisse expanded on the
accusation of indolence:

> How can we prove to people that they are lazy and that they are not actually walking? [. . .] All I can do myself here is to return the summons to you and say: Show us your studies. Until you have provided this demonstration, one is justified in accepting the fact signaled by public notoriety that accuses you of making poor use of your time. [. . .] Can one reasonably sustain that the school of Montpellier has, with its most recent *maîtres*, that is to say for more than a quarter of a century, actively worked toward progress in practical medicine? [. . .] Is it not all too certain on the contrary [. . .] that she has little by little fallen into a state of languidness that leaves one sometimes wondering if she is dead or alive? [. . .] You watch science pass before you as if it were a spectacle [. . .] Your scientific spirit resembles somewhat your medicine: you inquire into everything, but you do not become directly involved in anything (55–56).

Finally—it is now February 1843—Peisse takes on the vitalist dogma of the medical school of Montpellier:

> And to begin I willingly accept your interpretation of the word dogma. [. . .] I would observe all the same that it is more usually employed in the theological sciences than in the natural sciences, and that its application to propositions in medicine is not at all the custom except in Montpellier [. . .] The vitalism of Barthez is in fact a natural offspring of Stahlianism; however much it may seek to deny its father, to disguise itself, to change its country and language in order to hide its origins, one will always recognize it. [. . .] The doctrine of Montpellier was born in primitive fashion as a reaction to the schools of iatro-mathematics and iatro-chemistry [. . .], of which Stahl was the principal author and advocate [. . .] Protests were raised quite early, it is true, against pure animism, but even while making an effort by *esprit de système* to separate itself, it incessantly falls back again [. . .] Yet I would inquire of you, what does this [vital] principle—which *feels*, which *perceives*, which *is determined* spontaneously, which *moves* with efficacy, which has *affections, appetites, ideas*—lack in order to embody the irrational spirit of Stahl? Evidently, Sir, it only lacks the consent of Barthez and yours (86–88).

In 1842 Lordat responds to all of these critiques in a short opuscule entitled *Apologie de l'école médicale de Montpellier*. Before attacking the preconceived notions of his adversary, one by one, he declares:

> [The school of Montpellier] is well aware that wrong-doers wish to drown her; she trembled that they could find your honorable and imposing signature on those calumnious and malicious acts of notoriety and rage [...] In condensing as far as I can the reproaches

that you make of our School, I believe that I may unite them under the following headings: −1° She has the spirit and the tendencies of the Platonic School, and as a consequence must be more contemplative than active. −2° She must be indolent because the Platonic spirit is indolent. −3° As a consequence, she is immobile. −4° She trains Physician–Philosophers rather than Philosopher–Physicians. −5° She is persuaded that the Medical Science is finished and has nothing more to achieve. −6° In everything she is sterile [. . . However] the School of Montpellier has preserved herself from a philosophical vice into which others have fallen; I am referring to Formal and Systematic Materialism, a hypothesis whose introduction into Medicine is as damaging in practice as it is absurd in theory (6, 10).

Peisse refuses to participate in the "secret negotiations" proposed by Lordat (1842) to reach a compromise (72). Lordat's adopted son Marcel-Henri Kuhnholtz (1794–1878) took up the polemic on his father's behalf in *Paris et Montpellier sous le rapport de la philososphie médicale* (1843), challenging the translator of Hamilton and his perfidious Parisian advisors:

We ourselves see very well that Monsieur Peisse is setting himself up as a *Medical Oracle* . . . But on what authority?—We still do not know.—We see very well that he is contriving to turn himself into a *Dispenser of all medical glories* . . . But for what motives?—We hardly know [. . .] How can he not realize that, not being a Physician in name or in fact, he is walking more or less at random [. . .] Monsieur Peisse told us himself, 'that there is no need for him to travel' in order to know our school: nothing obliges him certainly, and he does well to stay at home, in his own surgery, if that is his preference. [. . .] It follows that his studies in medicine have been carried out, a bit like the voyage of Monsieur de Maistre, *autour de sa chambre* [. . .].

[With regard to Peisse's quotations from Lordat's writings] But why falsify a text? Why ridicule a quotation? Why seek to present Barthez here as a fool? . . . All that is *Feuilletonism*: it is necessary at all costs to appear to be in the right . . . ! (13, 22).

Kuhnholtz (1843) then turns to the experiments in physiology that were being pursued in Paris and the questionable results that were obtained by the organicians:

Yes we are agreed, the Doctrine of Montpellier, delivered in a very small number of books, has not produced great developments except in Oral Teaching [. . .] [But] what remains for Science of the hundreds of thousands of volumes that Broussais and the *Physiological Doctrine* have realized during the course of nearly twenty years? [. . .] (2) Let us therefore follow Monsieur Peisse in his examination of the question of Pathology [. . .] Certain Authors, of this doctrinal color [Organicism], have wanted to make us believe that Delirium is always linked to the Inflammation of some encephalic organ, and above all to the Arachnoid; but a great number of facts prove the contrary [. . .] (3) It is said, relative to the Paralysis of the Tongue: 'Monsieur Foville explains it as a sign of an affliction of the Ammon's horn [Hippocampus]; Monsieur Bouillaud regards it as closely linked to a lesion of the anterior part of the hemispheres or of the anterior lobe . . .': would it not be just as valid if, in their excess of perspicuity, these two observers had designated one as the North Pole, the other as the South Pole! [. . .] (4). [In] Paris, Medical Physiology leads directly to the adoption of the fine Principles that M.J.A.X.,[6] a Member of the Royal Academy of Medicine, [. . .] had the temerity to proclaim and publish in a book entitled *De l'épicurisme et de ses applications* [. . .] Here is what the Principles of the Author can be reduced to: *A.* [. . .]; *B.* There is no God; *C.* There is no soul. *D.* The belief in a superior Being and in the Immortality of the Soul, is eminently pernicious to Society [. . .] The Medical Philosophy of Montpellier, if she were ever placed under the obligation to express herself clearly on what concerns Morality and Theology, would believe herself under obligation to adopt, on these four points, diametrically opposed ideas (17–19, 54–55, 57, 65–66).

There followed a commentary in which Kuhnholtz declared that he felt called upon to quote this passage, as he had already done in many other publications, in order to wither "the revolting cynicism of which it is the expression". Thus, scientific rivalry gave way to sarcasm and—what is even more interesting—spilled over into the question of the existence of the soul and of God, which do fall within the official thesis of vitalism defended by Paul-Joseph Barthez.

The quarrels did not stop here. One finds further echoes of them in the *Protestation* by Alexis Alquié (1845); in *Retour vers l'hippocratisme* by Barbaste, who repeated that "The [Parisien] innovators require a philosophy that would be a *little too dirty, a too little crude* to accord with the tastes of Madame de Sévigné" (1852); in the exchanges between Jean Sales-Girons of the *Revue médicale* and Louis Saurel of the *Revue thérapeutique du Midi* (1852); and in the letter that Chrestien sent on July 10, 1860, to Rouland, Minister for Public Education and Religion, in which he recast the irreconcilable antagonists in this story as the figures in a Greek tragedy, with the city of Paris as the soothsayer Calchaas and Montpellier as the heroine Iphigenea . . .

2. The Prestige of the School of Montpellier

The history of medicine in the nineteenth century shows that the climax of the controversy between vitalism and organicism coincided with the most productive period of medical research ever achieved in France. Utilizing the revised data of Garrison (1992), Ben-David (1994) established that in the period between 1800 and 1830 medical discoveries in France surpassed those made in England, Germany, and America (Table 1).

Toward 1840 Germany would overtake France, before ceding primacy to the United States beginning in 1910. This coincidence leads to the postulation that the controversy between the schools of medicine of Paris and Montpellier may have been intimately linked to the level of productivity attained by France during the first half of the nineteenth century.

While a relativist picture of the controversy can be easily reconstructed on the basis of testimony and exchanges between the physicians from the two schools, the sociological studies of Ben-David (1994) offer a framework for the development of alternative, rationalist hypotheses. Ben-David dedicated himself to a comparative analysis of the scientific and medical organizations in different countries during the nineteenth century. He showed that differences in scientific productivity cannot be interpreted as the direct consequence of the comparative wealth of different nations nor of the differential circulation of knowledge in these countries.

A great number of reasons have been offered to explain the leadership of Germany over France beginning in 1850: its aptitude in identifying new fields of research; better organization within the university system; a high level of specialization in teaching; the existence of numerous institutions that facilitated the mobility of scientists. Ben-David (1994) weighs the importance of these factors and shows that experimental medicine in Germany was actually based in part

Table 1. Number of medical discoveries by country (1800–49).

Decade	France	England	Germany	U.S.A.	Rest of the world
1800–1809	9	8	5	2	27
1810–1819	19	14	6	3	47
1820–1829	26	12	12	1	57
1830–1839	18	20	25	4	71
1840–1849	13	14	28	6	68

on the French model and that the *Habilitationsschriften*, despite its level of specialization, was often comparable to the French *agrégation* (115). According to Ben-David, there was only one institutional factor that could explain Germany's ascendancy—the autonomy of its universities. This contrasted with the centralization of medicine in France, which by 1850 was no longer able to meet the requirements of medical research. It would appear that the centralization encouraged by the Montagnard party within the Convention (1793–94) during the French Revolution and further reinforced by the Premier Consul Napoleon Bonaparte in the Year VIII (according to the Revolutionary calendar, corresponding to 1799) in the long term slowed the autonomous development of professorships in specialized areas[7] and constrained scientists to engage in medical research as a secondary—and often dilettante—activity. Thus, at a time when centralization in France was preventing the multiplication of institutions of research, the Prussian government was proceeding with the creation of more and more new laboratories (Ben-David 1994, 123). Equally, the mobility of the scientist in France within his own institution was being reduced and all ambition suffocated by its system of mandarins.

While accurate in its broad outlines, this portrait does not take into account the fact that there was a certain amount of diversity in the academic institutions in France. Therefore—within the framework of this study—a more nuanced picture must be drawn that includes a comparison of the differing situations of the schools of medicine in Paris and Montpellier. The prestige of the faculty in Montpellier at the end of the eighteenth century and the beginning of the nineteenth century can be gauged qualitatively on the basis of a certain amount of historical evidence. It can also be measured using more tangible indices, including the number of students enrolled and their access to sources of knowledge.

1. Student Enrollment

Research by Julia and Revel (1989) documents the asymmetry in the student population at the faculties of Montpellier and Paris during the Ancien Régime. The numbers of first-year registrations (*matricules*) at the two universities testify to the net preeminence of Montpellier (Table 2).

The total numbers of registrations (*matricule, baccalauréat, licence et doctorat*) can be reconstructed starting with the official registers and the university fees levied (Julia and Revel 1989, 472–473, 478–479).

Table 2. Number of first-year students enrolled at the schools of medicine in Montpellier, Paris and Strasbourg (1753–89).

University	Period	Students
Montpellier	1760–89	71.2
Paris	1753–74	45.5
Strasbourg	1760–89	35.7

They confirm the primacy of the school of Montpellier during the period 1780–89, the average number of students per year being 268.4 at Montpellier and 106.6 at Paris.

Another clue lies in the results of an inquiry into the health professions launched in the month of Prairial, Year XIII (June 1805) by the Ministry of the Interior. The number of physicians who had graduated before 1794 from the schools of medicine in France and who were exercising their profession in the period 1803–06 shows an even more striking disproportion between Montpellier and the other faculties (Table 3).

Julia (1992) justly points out that in weighing this data we must take into consideration two factors concerning the schools in Strasbourg and Paris. First of all, a considerable number of the students enrolled in Strasbourg came from other countries, principally Switzerland and Germany, and would not remain in France after completing their degrees. Secondly, the fees demanded by the faculty in Paris were so

Table 3. Data showing where French physicians practicing during the period 1803–06 had studied for their medical degrees (Julia and Revel, 1989).

University	Headcount	%
Montpellier	1101	45.9
Toulouse	226	9.4
Caen	195	8.1
Reims	156	6.5
Besançon	122	5.1
Douai	96	4.0
Nancy	95	3.9
Paris	72	3.0
Strasbourg	38	1.6
Other	297	12.5
France	2398	100.0

exorbitant (on average 4,300 livres) that entrance was effectively reserved for students from well-to-do families (Julia and Revel 1989 281). Those of more modest means followed nearly the entire curriculum in Paris before matriculating at a less prestigious university in order to obtain their diplomas. They tended to choose one of the schools closest to Paris, such as Reims, Douai, or Caen. Even so, the diplomas conferred by Montpellier were two times more numerous than those granted by these other four universities combined (Table 4).

Conclusion: before the closure of the faculties voted by the revolutionary Convention in Year I (1792), one physician in two had studied at the school of medicine in Montpellier, while only one in five graduated from the faculty in Paris.

2. Access to Knowledge

The first hypothesis that comes to mind when seeking to explain the fame of Montpellier's school of medicine is that it had succeeded in conserving its prestige as one of the most important centers of learning in Europe since its founding in the Middle Ages. This hypothesis faces all the periods of crisis that the faculty of medicine had to face over the course of the centuries. There are more cogent reasons to explain her far-reaching influence in the years immediately preceding the controversy. It appears that Montpellier succeeded in creating an exceptional medical library, a crucial font of knowledge for its professors and students. Gabriel Prunelle, whose nomination as librarian of Montpellier's (newly named) school of health was approved by the Ministry of the Interior on 28 Vendémiaire of the Year VI (1797), would be the moving spirit behind this effort. In the Year XII (1803), as he began to encounter difficulties obtaining the books that he retained necessary for his library from the national depositories, Citizen Prunelle requested the support of Jean-Antoine Chaptal, then Minister of the Interior, who had received his degree in medicine from

Table 4. Number of medical doctors from Montpellier in comparison to the constellation of Paris–Reims–Douai–Caen, and other provincial universities.

University	Headcount	%
Montpellier	1101	45.9
Paris-Reims-Douai-Caen	519	21.6
Other	778	32.5
France	2398	100.0

Montpellier in 1777. Seconded by many members of the Académie des Inscriptions et Belles-Lettres, Chaptal appointed Prunelle "commissaire du Gouvernement chargé de l'inspection des bibliothèques et dépôts littéraires" (Vidal 1958, 90). This allowed him to gain access to the book depositories of Mans, Chartres, Dijon, Auxerre, and Troyes.

An examination of the correspondence during the period 1802 to 1807 between Prunelle and Gaspard-Jean René, the director of the school of health in Montpellier testifies to the extent of his bibliographic discoveries (Vidal 1958). Citizen Prunelle obtained the right to participate in sales and auctions during which he acquired important private collections such as those of Cardinal Albani and Charles-Louis L'Héritier. Prunelle wrote that he wished to "procure for the library those modern books of medicine of which we are absolutely devoid; I requested 10,000 Francs for this objective," and indeed he was so successful in his mission that the premises at Montpellier rapidly became insufficient to house the library's new collections. Prunelle did not limit himself to prospecting for books in France. He established ties with Hannover, and suggested to René that he contact the faculties of Leiden, Göttingen, and Jena in order to establish a regular exchange of university theses (Vidal 1958, 91). What is more, when Paul-Joseph Barthez, personal physician to the emperor, passed away in 1806 he left his entire collection of some five thousand volumes to Montpellier in order "to augment its public library." Eventually Gabriel Prunelle returned to Montpellier where he was appointed Professor of Medicine and the History of Medicine by a decree of Napoleon on November 10, 1807. In his *Cours de propédeutique médicale*, he expressed his conviction that access to knowledge was one of the essential conditions for progress in all fields of knowledge:

> The most fundamental working tool, is a library in which one finds, not precious manuscripts, but a collection of all the works of value that the ordinary means of an individual will not permit him to acquire, and all the books of manuals that could serve for the use of the students [. . .] Messrs Students, the studies that you have completed to the present are, in a way, nothing other than a preparation for those that will occupy you for your entire lives [. . .] (*Des études du médecin*, November 17, 1815).

This program had already been launched by Charles-Louis Dumas, professor of medicine at Montpellier, when Prunelle was beginning to comb the depositories and archives of France for books to add to the university's library (20 Germinal, Year XII):

> It is necessary to enrich our libraries with all the books that attest to the progress of medicine [. . .] and to prepare for it the greatest resources for its future development [. . .] A certain soul of genius [an allusion to Barthez] envisaged the progress of the human spirit as limitless [. . .] (*Discours sur les progrès futurs de la science de l'homme*).

One can easily deduce from a study of the student population and its library of scientific texts that the school of medicine of Montpellier had, during the decade of the 1810s, the necessary conditions and sufficient resources to engage—on quite an equal footing—in a power struggle with Paris. Its student numbers surpassed those of Paris and the school had comprehensive holdings of works to draw upon in the development of her medical doctrine.

3. Sociohistorical Analysis

At this point it is appropriate to examine the motives for the controversy, and to interpret why the dispute was in the end settled in favor of the Parisian school of medicine. In fact, the failure of Montpellier to impose the doctrine of vitalism has always raised numerous questions. Can it be explained by a difference between the experimental and speculative methods? Between the doctrines of reductionism and holism? Between the philosophical presuppositions of materialism and spiritualism? Did the divide between Montpellier and Paris form reflect a difference in scientific productivity (high versus low), or the differing statutes of the institutions (nonuniversity versus university)? Finally, was there a collusion of professional interests (centralism versus regionalism) or political interests (democracy versus monarchy)? One might well conjecture that all of these elements played a role, and each hypothesis merits close examination.

The analysis based on external and internal factors that will be undertaken here might appear to some to involve an excessively detailed look at reality. It is true that the separation between institutional, professional, and political elements is never perfectly clear cut (for example, the organization of a scientific institution will always depend, at least in part, on political support). But this separation of factors appears unavoidable. As we have seen, even those who adhere to the SSK principle that science and society are inseparable sometimes do so (on this point, see Lécuyer 1988, 406). Therefore this type of analysis is perfectly legitimate and it remains to define the order in which we will proceed.

In the study which he consecrated to this turbulent period in the history of the school at Montpellier, Dulieu (1990) surmised that professional and political interests were factors in the controversy, but nonetheless maintained that the quarrel centered above all on points of medical doctrine:

> Two schools would in this way clash with each other continually: Paris and Montpellier. Let us say immediately that professional jealousies played a role but also political questions perpetuated by the numerous changes in regime that the nineteenth century witnessed. It nonetheless remains that the spirit in which the study of medicine was approached would constitute an essential element in these controversies (212).

Thus we are justified in beginning our analysis with questions of method and doctrine, gradually proceeding to an examination of extrascientific factors.

1. Internal Factors

1. The interpretation of the clinical signs. In 1819 Bérard published *Observations cliniques pour servir de preuve à la doctrine médicale de l'école de Montpellier*, in which one finds a description of the complications that developed in a case of nephritic colic. The patient, a 60-year-old magistrate, underwent an abdominal operation on April 24, 1819, during which "an oval kidney stone weighing six ounces" was removed. On the eighth day after the operation the patient suffered a bout of delirium. On the ninth day there was spontaneous bleeding from the wound with the loss of one and a half ounces of blood. On the tenth day fresh, bright red blood once again flowed freely from the wound. All of these events occurred at the same hour on different days. Here are the clinical signs on which Bérard bases his diagnosis:

> The weakening two days before, the slight hemorrhaging the day before, above all the coincidence of the hour in which these events took place, the pronounced spring-like conditions of the season, the coincidence of a fairly large number of intermittent fevers [in the patient . . .] came together and led us to believe that there was present an intermittent daily fever of which the hemorrhaging was only a symptom (1819, §1).

Bérard diagnosed an "affection générale" that was directed by a "génie périodique pernicieux" (*sic*). In his eyes this case provided clear proof of the doctrine of Montpellier, which postulated the existence of general (nonlocalized) ailments with no organic cause. It goes without saying that the patient's symptoms would have been interpreted very differently by the physicians in Paris and neither the coincidence in the timing of the events nor the season of year would have been taken into account in establishing the diagnosis. What the doctrine of vitalism accepted without hesitation as pertinent data, organicism rejected just as categorically. How could a doctor attribute the complications in this patient to the alteration in a vital force not localizable to a specific site in the organism and, what is more, immaterial?

This case demonstrates the diametrically opposed styles of reasoning which prevailed in Montpellier and Paris. Vitalism tended to unify the seat of all illnesses while differentiating between ailments (using classifications such as fevers and *éléments morbides*) while organicism collected the ailments and differentiated their point of involvement. This is just one example among many that shows the appreciable difference in conception between the two medical schools.

2. *The methods.* One of the central issues in the vitalism versus organicism controversy concerns the methodology that the protagonists thought should be applied to the study of medicine. The difference between the two schools could be summed up as the opposition between the experimental method and the speculative method.

The promoters of Parisian medicine were all clinicians and reputable clinical observers. Corvisart (1818), who practiced internal medicine at the Hôpital de la Charité, rejects all medical theory and claims that his *Essai sur les maladies et les lésions organiques du coeur* constitutes a "work of pure practice, founded on observation and experience" (iii). Laënnec as well, who had started his career in the unit run by Corvisart at the Hôpital de la Charité, took care not to embrace any abstract medical theory. The priority for him was direct clinical experience and this was no mere theoretical recommendation; in 1822 Laënnec conducted twenty-two autopsies during the months of September and October alone (Ackerknecht 1986, 122). It is telling that he adopted as his motto *Ars medica tota in observationibus*. Here are the concluding words of his doctoral dissertation:

> I belong neither to the Ancients, nor to the Moderns: I follow both when they speak the truth. I believe in frequently repeated experience (1804, 38).

The same orientation can be seen in Broussais, although he did not contribute, strictly speaking, to the birth of experimental medicine. Before being appointed to a professorship at the Hôpital du Val-de-Grâce (1814) Broussais served as a surgeon with the French navy. As a doctor in the Army of the Ocean Coasts and in the Napoleonic wars, he profited from a vast store of clinical experience to develop his localist theses. Finally, there is no need to include a detailed study of François Magendie's and Claude Bernard's works, which stand out as among the most significant contributions of French medicine to the domain of experimental physiology and pathology (Ackerknecht 1986, 161).

The work of Marie Francois Xavier Bichat is worth mentioning here, for his metholodology had many points in common with that of the Parisians. He did not belong to the school of Paris, having studied and practiced in Lyon, and indeed some of his theoretical writings have led him to be described as a vitalist. Due to the revolutionary unrest in Lyon, he was forced to flee and took refuge in Paris in 1793. A brilliant physician, he began teaching a course in anatomy in 1797, which included dissections and demonstrations in physiology. He carried out more than six hundred autopsies during the winter of the Year X (1801), which led to important discoveries regarding the nervous system, such as the function of *contractility* in the voluntary nervous system and what he referred to as *sensitivity* in the sympathetic nervous system. Moreover, Bichat's approach was often experimental in nature and some of his clinical descriptions were based on *in vivo* studies, for example of arterial blood transfusions or the absorption of air at the level of the intestine (Ackerknecht, 1986). As Dulieu (1975) writes, "Bichat, who did not study at Montpellier but whose father was a doctor there, would not deny the existence of the soul but, this not having been subjected to experiments, he placed it to one side, preferring to occupy himself with what he could observe" (213).

The doctors in Montpellier took precisely the opposite approach, being little inclined to practice either observational medicine or experimental medicine. For his entire life Barthez, who received his degree in 1753, taught a theoretical form of medicine that was completely detached from observation and practice. The only clinical study he ever published was a treatise on gout (1802). The case of Dumas was more nuanced, because he compiled an admirably thorough anatomy of the human muscular system (1797), a study of chronic diseases (1812), and, most significantly, an *Introduction à la science expérimentale* (1800), in which he defended an inductive approach to medicine, although he was more interested in the theory than its application. Lordat presents a similar profile to that of Dumas. A dogmatic theoretician, he contrived somewhat tardily a justification for the doctrine of Montpellier by linking it to an inductive experimental philosophy (1848, 1858). But this was an abstract recommendation not supported by any concrete studies in experimental physiology. Bérard as well did not participate in any of the medical discoveries of his time; his only contributions were an essay on variola (1818) and some clinical observations intended to "serve as evidence" of the doctrine of Montpellier (1819). Only Alexis Alquié, head surgeon at the hospital Hôtel-Dieu in Montpellier, set out to study specific problems in clinical surgery and clinical pathology, without for all that renouncing the doctrine of Montpellier.

Hence, one can observe a clear asymmetry between Paris and Montpellier, between the advocates of experimental observation and the partisans of an anthropological doctrine that was more speculative than descriptive. In addition, the more experimental physiology progressed, the more deeply "Montpellier entrenched itself over the Barthezian doctrine" (Lavabre-Bertrand 1992, 63). Therefore, the speculative versus the experimentalist style was the first cause of the disaccord between the two schools.

3. Doctrine. The medical controversy had more fundamental repercussions on the points of doctrine most directly connected with these two methods. Observation and experimentation were incompatible with the vitalists' global approach to the human body, but they played a crucial role in the development of the localist and organicist doctrines in Paris. The latter—which sprang in part out of the theory of irritability proposed by Haller and by Tommasini—questioned the Hippocratic concept of "general disease" according to which all local ailments originate from a general imbalance in the human body. The localist school of thought would be all the more vigorously defended as there were in Paris at the time a number of adherents to the animism of Stahl, which proposed, by a sort of extrapolation of Hippocratic medicine, that the soul, as the seat of all illnesses, should be left to its own resources to restore the body to a state of equilibrium and health.

In contrast, according to the doctrine of localism every ailment is the result of a specific organic dysfunction. This can be recognized as the medical equivalent of the thesis known as reductionism: the body does not consist of an indivisible whole, but the ordered sum of different organs and functions. The localists applied this thesis to a multiplicity of subjects. Broussais, for example, maintains that states of consciousness could be reduced to the product of the action of an "internal nervous substance" (1826). Corvisart was not a localist *stricto sensu*, but one finds in him the same insistence that man should be viewed as a mechanism whose various parts could be studied separately. The project *Des localisations et des causes des maladies étudiées au moyen des signes diagnostiques et confirmées par l'autopsie* suffices to distinguish Corvisart's thinking from the school of Hippocrates, although when it came to the actual practice of medicine he adopted a more nuanced approach. His student Laënnec recounts that Corvisart frequently praised the—passivist—methods of Hippocrates, while at the same time not hesitating to utilize active therapies, such as cauterization in cases of phthisis (Ackerknecht 1986, 173).

While the experimental method was compatible with the reductionist theory of medicine inaugurated by Corvisart and Broussais in Paris, the speculative method of Montpellier required a holistic approach, as indeed was overtly suggested in certain texts by the advocates of vitalism. In a lecture delivered on 20 Germinal, Year XII (1803), Dumas, who was a professor in Montpellier, laid out the nature of the speculations engaged in by the school in its pursuit of an all-embracing, holistic form of medicine:

> Why separate philosophy and medicine? Do they not have as their object, one and the other, to reassemble all the reports that link man to nature? [. . .] Medicine is composed of a vast stock, inexhaustible in receiving the tribute of many sciences, of which it has become the fortunate depository. The destiny of a prodigious multitude of knowledge rests therefore in its hands [. . .] Barthez sees all the sciences at the service of medicine and all-powerful medicine at the service of man [. . .] (*Discours sur les progrès futurs de la science de l'homme*).

This holistic conception was – allowing for a few semantic nuances – in keeping with various prestigious antecedents. Parts were coherent with the canons of ancient medicine and, every time they were faulted, the vitalists of Montpellier were quick to claim their connection with Hippocratic doctrine.[8] But this was not the Hippocratism of physicians such as Sydenham, Barbeyrac, Corvisart or Laënnec, who saw in the Greek physician the founder of a medicine based on observation. Hippocrates was venerated in Montpellier as the author of the maxim: "Carry philosophy into medicine and medicine into philosophy." Albeit congruent with the Hippocratic notion of a *natura medicatrix* that regulates and reequilibrates its own dysfunctions, this was a somewhat original reading that can be found in various texts from the *Discours sur le génie d'Hippocrate* by Barthez (1801) to the essays of Lordat, *Vicissitudes de l'anthropologie hippocratique* (1856) and *Rappel des principes doctrinaux de la constitution de l'homme énoncés par Hippocrate* (1857). They all bear the hallmarks of a holistic conception of medicine, which Barthez referred to as "the Science of Man" (1778) and which Lordat would later call "Anthropology" (1852). For Barthez and his disciples, the "vital principle" was everywhere and acted on every site in the body. It was a diffuse force that guaranteed the indivisible unity of the human body. Alquié (1846), for example, recommended the study of man in his totality, without allowing oneself to be captivated by local manifestations (133, 217).

This constitutes the second fundamental divergence between the doctrines of Paris and Montpellier (even if one can always detect a few disparities in viewpoint among members of the same school).

2. External Factors

1. Scientific productivity. To begin with, let us focus on the works published by the participants in this controversy (Appendix 4). One can gauge the balance of power between the two schools by comparing the number of books and papers that were aligned with the doctrine of Montpellier or the doctrine of Paris.[9] This can provide us with an approximate measure of: 1) the scientific productivity (p: number of publications); 2) the scientific credibility (r: number of reprints and subsequent editions published); and 3) the overall volume of activity (t: total number of publications and subsequent editions) (Table 5) of the two schools.

One notes immediately that the productivity of the physicians in Paris was superior to that of Montpellier (p: seventy-seven vs. forty-three), that the recognition accorded their works was higher (r: fifty-seven vs. fifteen), and that the total volume of publications testifies to the overwhelming superiority of the Parisian school (t: 134 vs. fifty-eight).

The level of scientific activity of the two schools can measured by determining the number of publications produced by the two schools decade by decade (Figure 2). This shows that the success of experimental and observation-based medicine was affirmed during the decade 1820 to 1830, and that the tardy response of Montpellier (1850–60) was insufficient to check this process.

2. The philosophical presuppositions. Some have interpreted this controversy as the confrontation between a materialist philosophy and a spiritualist form of metaphysics, one that extended beyond

115

Table 5. The number of works published by physicians who supported the doctrine of vitalism, organicism, or some elements of both (1790–1870).

Author	Vitalists			Organo-vitalists			Organicists		
	p	r	t	p	r	t	p	r	t
Barthez	7	9	16						
Dumas	6	3	9						
Lordat	15	0	15						
Bérard	7	2	9						
Alquié	8	1	9						
Bichat				5	19	24			
Dugès				4	0	4			
Corvisart							4	2	6
Broussais							20	18	38
Laënnec							5	6	11
Magendie							25	14	39
Bernard							23	17	40
Total	43	15	58	9	19	28	77	57	134

p: publications; r: reprints, subsequent editions, and posthumous works; t: total.

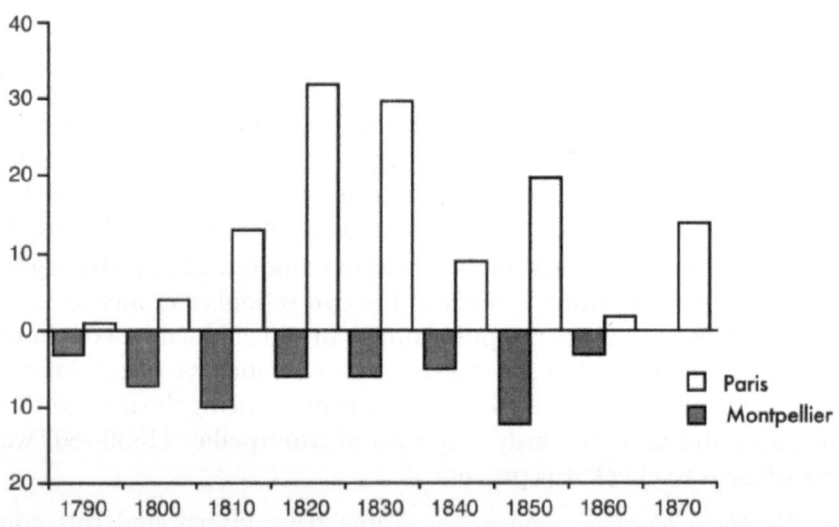

Figure 2. Total number of medical works published by vitalists and organicists (1790–1870).

the science of medicine to broader speculation on the relationship between man and reality.

Corvisart presented himself openly as an adversary of speculative theories, although he adopted some of Bichat's theories on cardiac activity and inflammation. Regardless of these borrowings, as Ackerknecht (1986) underlines, "Corvisart was not a vitalist; for him man was no more than a machine" (111). Concrete knowledge of vital phenomena was worth more than any speculation on the origins of life. This conviction was shared by Broussais, although for his part he would construct an entire system based on this position. Broussais remained a staunch defender of the materialism of Cabanis for his entire life, and even more stubbornly after noting a diminution in his audience beginning in the 1820s. In the essay *De l'irritation et de la folie* (1826), he attacked the spiritualist and eclectic philosophies of the period, stating that the practice of medicine should detach itself completely from any speculation on the soul as the seat from which the vital functions were coordinated.

The doctrine of Barthez was on this point diametrically opposed to the thesis promulgated in Paris, because the "vital principle" was based on just such a form of coordination. All the same, Barthez did not support the notion of animism. At the end of the eighteenth century, vitalism was presented as a doctrine that reconciled the irritability theory held by Albert von Haller (1708–77), William Cullen (1712–90) and John Brown (1735–88) and the animism of Georg Ernst Stahl (1660–1734). The vital principle therefore occupied a position equidistant from the notion of the soul and the theory of physico-chemical phenomena as being responsible for illness (Dulieu 1983). An essential difference between animism and vitalism was that, contrary to the soul, the vital principle was ephemeral in nature. Consequently, as Lavabre-Bertrand (1992) observed, the opposition materialism versu spiritualism is not sufficient to characterize the issues at stake in this controversy. The doctrine of Barthez does not presuppose the extrapolation into the metaphysical realm that some historians and sociologists of the sciences have attempted to associate with it. Two elements assure us of this.

First, in his theoretical writings Barthez never attempted to describe the seat of the vital principle. Nor did he venture to speculate on the origin and nature of this entity. His vital principal emerged from an abstract synthesis of Bordeu's thesis of *petites vies*, that is, of the organs as so many separate "little lives." Barthez believed these little lives to be the expression of a unique principle that united them, coordinated them, and whose function was to keep the body alive.

Second, sociologists will be aware that the doctrine of Barthez exercised its influence on no less a figure than the philosopher Auguste Comte, who attended his lectures in 1816–17 when he returned to his home town, Montpellier, after his dismissal from the École Polytechnique. One would run up against some contradictions therefore if one sought to impose a link between Parisian medicine and positivism while ignoring the connection between Comte and Barthez.

Clearly vitalism was opposed to the reductionism of experimental medicine, but this did not mean that it posited a spiritualist form of metaphysics. Barthez's doctrine remained a phenomenology. Any discussion of spiritualism is only pertinent in relation to the successors to Barthez in Montpellier. The vitalism embraced by Bérard, in the form of a *Doctrine des rapports du physique et du moral pour servir de fondement à la métaphysique* (1823), denotes linkages of the doctrine with the animist thesis. The introduction of an element of spiritualism is even more explicit in the writings of Lordat. Alongside organic forces, Lordat proposes the existence of a higher order of inorganic forces that make up the human soul. This combination of the thinking soul with a vital force lay at the origin of his thesis of the "duality of human dynamism" (1854). Lordat would go so far as to place, behind the immaterial soul, a "soul of second majesty" that presided over vital phenomena (Dulieu 1983). This was a translation of the vital principle, with connotations on the political as well as the religious plane. But Lordat's mingling of metaphysical and doctrinal questions remained the exception.

The tardy intransigence of the school of Montpellier over a philosophical point of a spiritualist nature is not sufficient, however, to explain the origins of the controversy. This had already begun to gain momentum at the time of Barthez, well before the spiritualist coda came to be grafted onto the doctrine of vitalism.

3. *The institutional framework.* As Schweber (1997) observes in a critique of the works claiming to be in the vanguard of the sociology of the sciences, "A second, more serious problem in the greater part of SSK studies concerns the near absence of any institutional considerations [. . .] The studies on controversies, for example, [pass] directly from technical interests to ideological and social interests, neglecting the influence of factors such as the formal organizations [. . .]" (89).

The role of institutional factors certainly merits more detailed examination.

It can be immediately seen that the two medical approaches spread via different institutional and scientific networks. There are two parameters that allow us to establish this: the position at the Academy of Sciences (*titulaire, adjoint, associé, correspondant*) of the various actors in the controversy; and their affiliation with either a university or a nonuniversity institution.

Apart from Broussais, who was elected to the Académie des Sciences morales et politiques (1823), three of the five Parisians were *membres titulaires* (full members) of the Academy of Sciences: Corvisart (1811), Magendie (1821) and Bernard (1854), the last of whom carried off the annual prize in physiology three times for his work on nerve tissue, in particular the sympathetic nervous system. In contrast, Montpellier did not count a single *membre titulaire* of the Academy of Sciences. Barthez and Dumas were associate members (elected in 1782 and 1800, respectively) while Dugès was a corresponding member (1836). This asymmetry arising from a source other than the merits of Barthez's thesis meant that the physicians in Montpellier suffered from a lower degree of scientific credibility. However, according to the relativists this asymmetry in scientific esteem could be interpreted as proof that the physicians in Paris benefited from an unfair advantage in terms of their access to the institute.

With regard to the division between university and nonuniversity institutions, the vitalist doctrine developed almost exclusively within the medical faculties, which were transformed into schools of health during the Revolution (between 1794 and 1806). The Montpellierians were all academics. During the Ancien Régime, Barthez had occupied the chair of anatomy and botany (1761). He was excluded when the medical school (for reasons that we will see below) was replaced by the school of health, but he was eventually nominated honorary professor in the month of Nivôse, Year IX (1800). In contrast Dumas was appointed to a chair of anatomy and physiology as soon as the school of health was created (1794). Later he would be assigned the chair in inpatient medicine (1809). Lordat held the chair of operative medicine and rare cases (1811) before being transferred at his own request to anatomy and physiology (1813), which in 1824 became the chair of human physiology when it was decided to divide the subject into two separate disciplines. Bérard became professor of therapeutic and medical matters (1827), while Alquié was appointed to the chair of outpatient medicine (1850).

119

In contrast, the organicist doctrine would be developed in Paris primarily by physicians working in institutions not associated with the faculty of medicine. Corvisart, who began as a professor of inpatient medicine at the Hôpital de la Charité, was nominated to the chair of medicine at the Collège de France—an autonomous establishment of higher learning—in 1797. Laënnec, after working at Necker Hospital, succeeded to the chair held by Corvisart at the Collège de France in 1822. Magendie as well was professor of experimental medicine at the Collège de France (1830–55). After his death Claude Bernard, who had been his assistant, would succeed to Magendie's chair (1855). It may be noted that when Bernard left the Collège de France it was in order to take up a teaching post in comparative physiology at another autonomous scientific institution, the Museum of Natural History (1868).

Therefore, while all of Barthez's disciples formed part of the faculty of medicine in Montpellier, four of the five Parisians engaged in the debate taught at institutions outside the university: four at the Collège de France and one at the Museum of Natural History (Broussais' situation, as we will see below, was atypical). The Collège and the Museum, both of which were historic institutions,[10] shared an important feature; their attachment to doctrine was less strong than at the university, and they viewed their vocation quite differently. Teaching did not consist solely of authoritative *maîtres* imparting knowledge to a submissive audience; professors were encouraged to conduct research and to pass on the knowledge gained from their latest discoveries to their students. In contrast, by statute the schools of health were not destined to be centers of research. From their inception their primary objective was to train students in the practice of medicine. As Ackerknecht (1986) reminds us, "The school of health saw itself as a specialized school, destined essentially for the formation of practicing physicians. Scientists were carefully relegated to 'sanctuaries' such as the Museum and the Collège de France, which were institutions intended essentially for research" (53).

Here one can observe the close correspondence between the institutional affiliations of the protagonists in this controversy and what we have been able to determine regarding their methods, doctrines, and scientific productivity.

4. Professional interests. After the decree of Year I, which led to the abolition of the universities, the revolutionary Convention proceeded, by means of a new decree issued on 14 Frimaire, Year III (December 4,

Table 6. Comparison of the numbers of students in the medical faculties of Paris, Montpellier and Strasbourg before and after the Revolution.

School of Health	Before 1789	After 1794	Variation
Paris	29.9 %	54.5 %	+82 %
Montpellier	46.7 %	27.3 %	−41 %
Strasbourg	23.4 %	18.2 %	−22 %
Total	100.0 %	100.0 %	

1794), to create three schools of health: one in Paris, one in Montpellier, and one in Strasbourg. This decree was accompanied by a measure which fixed the maximum number of students for each school: 300 in Paris, 150 in Montpellier, and 100 in Strasbourg. By comparing these numbers with those in Table 2, the change in the level of enrollment between the Ancien Régime and the Convention can be calculated (Table 6):

Montpellier suffered the most marked decline (41% compared to 22% for Strasbourg), while the number of students in Paris rose by 82%. Thus, a new hierarchy was imposed among the schools in a political decision that clearly reflected the process of centralization being promoted by the Montagnards. This decision sealed the fate of the faculty in Montpellier, which saw its influence diminish as it was reduced to what was in effect a provincial university. With its interests greatly damaged, the faculty of medicine in Montpellier had every reason to feel resentment against the authorities in the capital. Partisan professional interests therefore became a factor in the controversy, giving rise to a secondary dispute over the merits of centralization versus regionalism, and the question as to whether Montpellier was required to blindly obey the decrees sent down from the capital.

On certain points, nonetheless, Montpellier fought back and managed to resist the changes imposed by its rival. In 1814, Friar Elysée, the surgeon to Louis XVIII, proposed reforms to divide the disciplines of Surgery and Medicine and to re-establish the Barber-Surgeons' Company. While the faculty of medicine in Paris supported this measure, the professors in Montpellier rebelled against a proposal that they considered, with some justification, to be a step backward. The *querelle* lasted from 1814 to 1816, but the merits of the arguments presented by Montpellier in a *mémoire* to the public authorities in

1814 were acknowledged. All the same, the opposition of Montpellier was motivated by professional interest rather than a point of principle. In fact, the real objective of Friar Elysée's proposal was to restore the right of the Premier Chirurgien du Roi to issue lieutenancy patents and licenses for the preparation of remedies, two sources of revenue that had been suppressed by the revolutionary authorities. Montpellier refused to institute these tariffs and at its university the teaching of surgery remained tied to medicine.

But still, even if it is obvious that any reorganization of a scientific discipline will have a direct impact on the professional interests of its members, it is difficult to envisage a similar determination in the controversy here. That Montpellier should have resented its reduction in status in the Year III and clung tightly to its prestigious past is possible. But one cannot see why the school would have chosen—a priori—to adopt a thesis simply because it contradicted what was being taught in Paris, for this would have meant placing her professional reputation at risk in the eyes of the medical community. The humiliation of the faculty in Montpellier could explain the choice of a medical doctrine that differed from the one espoused in Paris, but does not explain why they chose to disagree specifically on the issue of vitalism.

5. *Political values and institutional affiliations.* Ackerknecht (1986) has shown that the principles of a medicine based on observation and experimentation—which, as we have seen, lay at the heart of the dispute between Paris and Montpellier—were fully subscribed to in the creation of the schools of health. This was a political decision, as is attested to by the report Count Fourcroy presents to the Convention on the 7 Frimaire, Year III (November 27, 1794):

> In the centralized Health School [. . .] practice, experimentation will be joined to theoretical precepts. The students will be trained to [conduct] clinical experiments, anatomical dissections, surgical operations, [and the use of] instruments. Reading little, watching much and doing much, such will be the basis of the new form of education that the *Comités* proposes to decree (47).

This text provides evidence of a possible relationship between institutional and political factors that we will now examine. In fact, one may ask whether the advocates and opponents of experimental medicine were not acting more out of political conviction than for scientific motives. Before making the overly hasty assumption that this controversy took the form of a rational debate, it is necessary first of all to determine whether the physicians in the capital might not have been more sensitive to the revolutionary movement than their colleagues in Montpellier.

A first clue in this regard is provided by the leader of the vitalist doctrine. It is known that Barthez was hostile to the ideals of the revolution. When the *Etats généraux* were inaugurated, he published a short work entitled *Libre discours sur la prérogative que doit avoir la noblesse dans la Constitution et les États généraux de la France* (1789). This treatise was doubtless motivated by the fact that Barthez had been serving since 1781 as the personal physician of the Duke of Orleans. When the opening events of the French Revolution exploded in 1789, Barthez judged that he was in a compromising position, resigned from his duties as professor at the school of Montpellier on 2 Thermidor, Year II (1793) and fled to Narbonne. If he was allowed to return in 1802, it can only have been under the protection of the Minister of the Interior Chaptal, who had studied at Montpellier.

But can one therefore conclude that all of the professors in Montpellier were royalists? Dumas, who was one of the first to take up the cause of vitalism promoted by Barthez, certainly does not fit this characterization. A professor at the École Centrale of Lyon, in September 1793 he appeared before a session of the *Assemblée du peuple* to demand that the city capitulate to the revolutionaries. The royalists dealt with him harshly and threw him into prison. Narrowly escaping execution, he obtained from Paris a transfer to Nice as a surgeon in what was to become the Army of Italy (the French army stationed on the Italian border). Upon his return to Montpellier, Dumas entered politics once again, siding with Cambacérès, a moderate who would nevertheless be exiled during the Bourbon Restoration. Dumas's political sympathies were finally rewarded when he was appointed the first director of the school of health in Montpellier (1794). He also served a brief tenure as the president of the municipality (1798–99).

Bérard constructed a more nuanced relationship with the political institutions of the Restoration. One of the decisions made by the Bourbons upon their return to power was to eliminate the locally-administered *concours* for professorships at the schools of medicine. Henceforth, professors were to be appointed by the authorities in Paris. And while Prunelle and Broussonnet were stripped of their functions due to their political antecedents, Bérard obtained a chair with the support of the Minister of Education and Grandmaster of the

University, Monseigneur de Frayssinous (1823). It is difficult to establish whether Bérard was motivated by political conviction or opportunism in the building of his alliances. This ambiguity is again accentuated by a perusal of his *Mémoire sur les avantages scientifiques et politiques du concours* published in 1820, five years after the Restoration, in which Bérard defended a position that was at the antipodes of the policy adopted by the Bourbons.

Jacques Lordat, who represented the last generation of adherents to the theses of the medical school of Montpellier, followed Barthez's example and sided with the Bourbons, but only at a relatively late date. When he arrived in Paris in 1793, the 20-year-old Lordat expressed his support for the new Constitution and then enrolled as an army surgeon (15 Germinal, Year II). It was well after he met Barthez in 1806 that Lordat began to defend a version of vitalism influenced by the conservative notions of Louis de Bonald (1754–1840). Around 1850, when vitalism was on its deathbed scientifically speaking, Lordat thought it opportune to propose an "accord" between his philosophical-medical conception—already strongly tinged with spiritualism—and the doctrine of Bonald, the leader of the Ultra-Royalists. We owe to Lordat two texts in a similar vein—*Accord de la doctrine de Montpellier avec ce que demandent les lois, la morale publique et les enseignements religieux* (1852) and *Apologie de la définition bonaldienne de l'homme* (1854)—that betray his monarchist and traditionalist convictions. He was then 81 years of age.

Little is known about the political opinions of Alexis Alquié, apart from the fact that he was a member of the commission of 1852 that had been formed to safeguard the interests of the faculty. One of its functions was to pledge the university's allegiance to the Second Empire, an act that in itself was only of limited political significance. It is well known that during the course of the nineteenth century university professors were regularly called upon to swear an oath of allegiance—sometimes to the Republic, sometimes to the Empire. This declaration of loyalty to the latest regime in place became a routine formality. On this occasion a request put forward by the commission was not, however, accepted; Napoleon III eliminated the local *concours* for professorships.

A similarly complex picture can be drawn of the scientific community in Paris during this tumultuous period, one in which we find revolutionaries, royalists, and many radical democrats who were attempting to modernize medicine while avoiding any entanglement in the political vicissitudes of the moment.

Broussais was without a doubt the French physician most deeply committed to the Revolution. Breton by birth and a liberal Bonapartist, he fought in the ranks of the Revolutionary Army against the Chouans (1792–94). In reprisal his parents were murdered at Christmas in the year 1795. Broussais accompanied Napoleon's army on its campaigns in Holland, Germany, Austria, Italy, and Spain. In 1828, when the university in Paris was dominated by the Bourbons and the Jesuits, Broussais published a polemical tract against the "Kanto-Platonicians," in other words spiritualists and other eclectic thinkers whom he went so far as to qualify as the "bootlickers of the Prussians." Here we find a reputable physician of the period highlighting the political overtones of vitalism. But we should not be misled by Broussais. He was a canny tactician and did not hesitate to play on the susceptibilities of the public to gain its support (Ackerknecht 1986, 82–83). It would be going too far to deduce from this a pure and simple determination of scientific content by political ideas. Broussais was making a calculated decision, raising the specter of the Prussian enemy in order to fill the lecture halls at a time when attendance was rapidly falling.

We have few indications as to the political ties of Corvisart. At the beginning of his career, under the Ancien Régime, he applied for a position at Necker Hospital. His application was turned down for what could have been political reasons: Corvisart refused to wear a wig. Another clue is the fact that he was appointed to a professorship soon after the reorganization of the medical schools, in 1794. An overt royalist would have had to wait for the Restoration before receiving such an appointment. At the same time his assistant Leroux des Tillets was dismissed in 1822 by the Bourbons.

In contrast, Laënnec maintained lifelong ties with the Jesuits and the royalists, a position that earned him the violent invective of Broussais. In Paris Laënnec frequented the circles of Mme. Récamier, Martinet, Antoine-Laurent Bayle, Cayol, and Chomel, all staunch reactionaries and Orléanists. However, it would be difficult to establish a direct connection between Laënnec's personal convictions and his scientific options. As Ackerknecht (1986) observes, "His traditionalist character was reflected in his politico-religious convictions, but is in total contradiction with the many innovations that he brought to medicine" (130).

This review of the political factors at play in nineteenth-century France shows that the originators of organicist medicine counted a liberal (Broussais), a moderate revolutionary (Corvisart), and an ally of the Royalists and Jesuits (Laënnec). The vitalist theses were defended by a monarchist (Barthez), a revolutionary (Dumas), and a man who embraced reactionary ideas toward the end of his life (Lordat). The other participants in the controversy do not appear to have had any marked political convictions. This dispersion of political opinions is consistent with the conclusions drawn by Schweber (1997) based on his statistical studies of nineteenth-century Europe, "In France the debates of 1837 and 1867 were characterized by a disjunction between different types of arguments [. . .] In other words, the positions of the individuals on technical questions do not permit us to predict their positions on political or professional questions" (85).

4. Conclusions

In this analysis of the controversy between the schools of medicine in Paris and Montpellier, if one accepts the premise that each of the factors discussed above played a relatively independent role vis-à-vis the others, it can be affirmed that only four factors show a close correlation with the divergence in position of the two schools— method, doctrine, scientific productivity, and institutional affiliation. The three other factors—professional interests, political values, and political affiliations—may have oriented the physicians in Montpellier toward a doctrine that was *different* from the one being taught in Paris, almost irrespective of the content of this doctrine. It remains to examine why relativist studies generally tend to arrive at the latter conclusion.

1. Productivity and Scientific Content

The controversy between Paris and Montpellier was marked by a singular historical feature that is difficult to reconcile with the principles of the relativist program, which presumes that the resolution of a scientific statement comes about through social negotiation. This chapter suggests quite a different conclusion: 1) The controversy was settled in favor of those who abandoned philosophical speculation for the practice of observation and experimentation; 2) The vitalist doctrine of Montpellier was vanquished by its Parisian opponents due to the more prolific research activities of the latter; 3) The primacy of the capital was supported by scientific institutions that were more open to innovative ideas in this period (the Collège de France and the Museum of Natural History).

The institutional factor—the only extrascientific factor that played a role in the controversy *stricto sensu*—does not provide an argument in favor of the relativist interpretation. This becomes obvious when one poses the question: Can one use the institutional factor to explain the *success* of the school in Paris? Contrary to all expectations, the answer is negative. Let us begin by reflecting on the mode in which a controversy may be settled. If a body of knowledge A cannot be subjected to testing—as in the case of the "génie périodique pernicieux"—it would be of little profit for the partisans of B, whose knowledge can be tested, to waste any effort in the refutation of A. The means required would be disproportionate to the ends that could be achieved, so there is every reason to limit oneself to the *implicit* rejection of A. The settlement of the controversy therefore will not proceed from an invalidation of thesis A, but from the level of scientific productivity of B and the accumulation of overwhelming evidence in its favor.

The relativists might then argue, based on the work of Ben-David (1994), that the productivity of the parties involved could have been determined by the social forms of organization in research. It is undeniable that independent institutions such as the Collège de France and the Museum of Natural History provided more favorable conditions for research in experimental medicine than the university: well-equipped laboratories, a respect for specialization, and less demanding teaching duties. Scientists could therefore dedicate themselves more fully to research. This is not to say that one should not extrapolate the analyses of Ben-David (1994) to the scientific *content* of the controversy. In fact, when Ben-David concludes that certain forms of organization may augment the scientific productivity of researchers, he does not attribute the success or failure of a scientific theory to the institutional structures. As Gad Freudenthal writes in his preface, "Social conditions may determine the 'how' and the 'when' of the emergence of new disciplines or scientific theories; equally, they may determine the rapidity of their development, but they do not have an impact on the *empirical validation* of scientific statements [. . .]" (in Ben-David 1994, 2). The institutional factor determined the productivity of the researchers but not the scientific content of organicism and vitalism.

2. The Failure of the Notion of Determination

In the end, the results of our analysis are consistent with the position that scientific content is subject to the criteria of rational proof. Therefore any simplistic schema positing the influence of society on science must be rejected, whether the modality be causal determination (Bloor 1976) or a network of allies as proposed by Latour (1987), "There is no great divide between minds, but only shorter and longer networks" (259) or by Shapin and Schaffer (1989), "He who has the most, and the most powerful, allies wins" (342).

The controversy that we have just examined suggests an entirely different conclusion, in particular because the divergences in the political views of the participants do not correlate with the differences in their scientific ideas. The argument of Collins (1981) is more admissible, to the extent that the relationship of efficient causality is replaced by a weak relationship between the scientific sphere and "political and social structures". In fact, the study of the controversy between Paris and Montpellier allows us to define this relationship more precisely. Our analysis shows that scientific activity depends on the social structures to which it is *most closely* tied, that is to say, the organizations within which the research is conducted. In this particular situation, the spirit of free inquiry at the Collège de France—which provided latitude for innovative norms of behavior—made possible the production of scientific content that was new and, *in this capacity alone*, opposed to the vitalist doctrine of Montpellier.

This relationship with an external factor does not signify that one is in a position to predict, given that it is an extrascientific factor, the content of the theory that is being influenced by it. The relationship here is neither causal nor functional; it belongs to the class of *significant* relations (Merton 1973, 12). The concept of *congruence*, limited in its meaning to the *harmonizing of the scientific content with the values and norms of scientific behavior*, is doubtless the concept with the most sociological pertinence to this type of relationship. All of these specifications reduce the applicability of the relativist approach, which does not distinguish between the different relationships that can exist between knowledge and its social environment; there is no reason to confound relationships such as causal determination (Boudon and Lazarsfeld 1966), functional interdependence (Merton 1965), structural homology (Panofsky 1967), and congruence (Merton 1973).

Finally, there are two formal arguments against the notion that a determination of the scientific content is at play in this particular controversy.

1) Let us suppose – for the sake of argument, adopting the assumption *ceteris paribus*—that the doctrine of vitalism was being defended by Paris rather than Montpellier. The thesis of the determination of scientific content by extrascientific factors would predict that the school in Montpellier would then have defended organicism, since the resentment caused by the decree issued in Year III, the professional interests at stake, and the institutional differences would have been divided in an identical manner as before. One might object that this abstract reconstruction is fanciful,[11] but it illustrates one of the inherent limitations of relativism. SSK is not capable of explaining why—continuing our argument *ceteris paribus*—the school in Montpellier could not have adhere to the doctrine of vitalism while carrying on its secondary quarrels with Paris over such matters as the number of students or how to restructure the medical profession . . . Why should a political or professional disaccord necessarily have an impact on the scientific content of a controversy? Such a relationship has been hypothesized in numerous studies, but no actual proof has ever been provided.

2) Alternatively, let us posit that Paris advocated an organicist doctrine, but we do not know what doctrine was embraced by Montpellier. The professional and political factors, taken in isolation, are still not capable of stipulating *a priori* the vitalist content of the doctrine of Montpellier. There is no valid reason for us to assume that the school of Montpellier would have defended vitalism rather than any other medical thesis of the eighteenth century that fell outside the canons of the new medicine—for example, the animism of Stahl or the humoralism of Friedrich Hoffmann to mention just two. If the relativists respond that Montpellier adopted vitalism because Barthez was teaching in that faculty, they must then—by the application of their own principle of symmetry—explain why medicine in Paris, which included savants who were quite open to the notion of vitalism such as Chaussier, Pinel, and Bichat, turned away from speculative medicine and dedicated themselves to a new approach based on experimentation. This is the second limitation of relativism, imposed by its conventionalist aspects.

The inconsistencies in the SSK argument revealed by this analysis lead us to support a rationalist approach to scientific knowledge. Even if such an approach grants that knowledge may be "constructed", its interpretation of the principle is different from the one proposed by the relativists. As Boudon noted (1994, 22), the mere fact that the discussion between opposing researchers may be protracted does not signify that the settlement of the dispute is *negotiated*. So long as neither side has at its disposition convincing proof, the discussion will continue. Therefore, the prolongation of a debate and the process of "the construction of knowledge" can be equally well interpreted within a rationalist framework by saying that a scientific discussion is concluded when rational and convincing proofs are advanced and understood by the opponents. Imre Lakatos (1980) has shown that researchers rarely abandon a theory if it has been contradicted only by one experiment conducted in isolation. But this does not in any way signify an arbitrary attachment to the contents. Lakatos explains this inertia by the "protective belt" that surrounds the "hard core" of a scientific theory.

A critique of SSK should not therefore call into question the idea of studying the scientific content of a controversy, but the *a priori* assumption that scientific activity necessarily implies a "social determination of the scientific content." The weaknesses of relativism that have been highlighted in this chapter prompt us to adopt two new principles:

—A sociological analysis must seek to distinguish the exact nature of relationships that the sociology of science tends to group together under the heading of causal determination, functional correspondence, homology, congruence, etc.

—It is necessary to recognize, rather than to deny in the name of a generalized principle of symmetry, the difference between the internal and external factors that can influence the production of scientific knowledge. Internal factors are much more important than external factors. Among the latter, however, the institutional factors directly tied to scientific activity have priority over factors with only an indirect association, such as politics and religion.

Notes

1. The controversy rose well before and its repercussions continued to extend for quite some time after these dates, but tensions between the two schools were never so high as in the period from the *Examen des doctrines médicales* by Broussais (1821) to the *Doctrine anthropologique de Montpellier* by Lordat (1852). During this period six books were published—in the years 1821, 1830, 1845, 1846, 1851, and 1852—whose titles referred directly to the doctrine of Montpellier. It was not until the appearance of Introduction à l'étude de la médecine expérimentale by Bernard (1865) that the leadership of the Parisian school was definitively consolidated.

2. In the first edition of *Nouveaux Élémens,* Barthez (1778, 26) declared that he sustained a theory of vitalism close to that of Medicus (1774). In the second edition of his book, he supported his case by citing Herder (1785, 108), who also made a distinction between the soul and the vital principle (Lavabre-Bertrand 1992a, 48).

3. It may be noted that the controversy between the schools of Paris and Montpellier had antecedents going back to the seventeenth century and the famous Antimony War. This dispute opposed Simon Courtaud and Isaac Carquet against the Parisians Gui Patin, Jean Riolan, and Charles Guillemeau. The falsehoods introduced by the *doyen* Courtaud in his apologia for the school in Montpellier unleashed the ire of the Parisians. Carquet took up Courtaud's defense in a *Seconde apologie* (1653). The Parisians countered with a *Responsio pro se ipso ad alteram Apologiam impudentissimi et importunissimi Curti Monspelcaniscellarii* (1654), and a *Defensio altera, adversus impias impuras et impudentes tum in se, tum in Principem Medicinae Scholam Parisiensem, Anonymi Copreae calimnias ac contumelias* (1655) and the dispute assumed an unusually intemperate tone (Dulieu 1983, 609).

4. "The painful circumstances in which the Faculty finds itself has committed us to push forward with the final version of our work. If one is to believe certain rumors that have come to light, the existence of the School of Montpellier is being threatened, and that of Paris could be granted a supremacy from the powers, which is nothing short of ridiculous, when it comes from any authority other than that of the superiority of talent" (Bérard 1819, 9).

5. Kühnholtz informs us that after leaving Montpellier, Peisse relied for most of his medical knowledge on private conversations that he had with Jean Miquel (1803–47) of the *Revue médicale.* Miquel as well had broken with the school of Montpellier after studying there until 1828 (Kühnholtz 1843, 21). We would note that the Revue médicale had no bias on this particular issue, since it counted among its collaborators Delpech, Dugès, Bérard and Prunelle of the faculty in Montpellier.

6. This is the pseudonym of Jacques-André Rochoux (1787–1852).

7. The restricted number of professorships led to serial nominations. To give just one example, Claude Bernard did not give up his chair in Experimental Physiology at the Sorbonne to Paul Bert until he was appointed to the chair of Magendie at the Collège de France (1855).

8. We can nonetheless cite a counterexample to this association between vitalism and Hippocraticism: Xavier Bichat, who was not an adversary of the vitalist theses, completely rejected Hippocraticism and any system of classification *a capite ad calcem.*

9. This statistical analysis only took into account medical works; inaugural addresses and historical notes that did not touch upon issues of doctrine were excluded.

10. The Collège Royal was founded in 1530 by François I; it would successively be transformed into the Collège National and then the Collège Impérial before becoming the Collège de France during the Restoration. The Jardin des Herbes Médicinales founded by Louis XIII in 1635 became the Museum

of Natural History in 1794, by the order of the Convention. These two institutions remained independent of the University of Paris.

11. This fantasy followed the ideas of Max Weber regarding the objective possibility of using "ideal pictures" (Phantasiebilder). As he wrote, "For all that, it is far from being 'idle' to raise the question as to what might have happened if, for example, Bismarck had not decided for war" (Weber 1964, 270).

4

Intromission versus Extramission in Oxford: An Essay on the Norms of Rationality

A crucial question in the debate between the relativists and the rationalists centers on the more or less rational nature of the conduct of science. According to the former, the norms of rationality are variable over time. Therefore, every period of history and each local context for the production of science may lead to a redefinition of truth and error, creating islets of knowledge that are incommensurable—in other words, "ethnotruths." For the latter, the norms of rationality that form the basis of science are universal or, at the very least, stable and independent of specific cultures. Consequently, there is no question of their being subject to periodic revision. Rather, they ensure the robustness and objectivity of scientific theories.

A multitude of arguments have been advanced on either side. The relativists insist that since all scientific production is *embedded* in its context, it is difficult to see how universal structures could impose themselves—everywhere and in every period—on the local processes of producing knowledge. One can only attempt to show how local scientific statements may become universal. The rationalists use the history of the sciences to show how cultures that are very different from one another may discover identical processes of reasoning or, alternatively, processes of reasoning that manage to arrive by different paths at identical results.[1]

This chapter will seek to show how the controversy that sprang up in the field of optics during the Middle Ages offers an exemplary case with which to test the SSK hypothesis that the norms of rationality are relative to the context of the production of science. The scientific statements under examination pertain to a specific branch of scientific knowledge—optics—at a specific time and place in the history of science—the *studium* of Oxford in the thirteenth century. The selection of this case for analysis is not fortuitous, because it presents two conditions that are quite interesting from a sociological point of view:

1) The extramission versus intromission debate individuates a major scientific controversy in the history of optics. This fact is in itself interesting for two reasons.

First of all, because the studies following in the wake of the Empirical Programme of Relativism (Collins 1981) have often endeavored to describe scientific controversies in such a way as to highlight "closure mechanisms" that depend, according to them, on social and political structures (Collins and Pinch 1994). The debate under scrutiny here could offer an occasion to verify this mechanism. Second, because any controversy raises a manifest set of arguments that may render visible the arsenal of reasons and paralogisms available to scholars at a given period in time (and which could remain implicit under other circumstances). This particular controversy began in antiquity, but was rekindled in the thirteenth century when scholars once again had access to authoritative texts, not only from ancient Greece and Rome (Aristotle, Euclid, Hero and Theon of Alexandria, Ptolemy, Damianus) but also from the medieval Arab world (Ibn al-Haytham, al-Kindī, Qustā ibn Lūqā, Ibn Sīnā, Ibn Rushd), texts which by this time were commensurable because they had all been translated into the same language—Latin. Among the Arab sources, one of the most important was *Kitāb al-manāzir (De aspectibus)* by Ibn al-Haytham (Alhacen) which treated a great number of questions with a clarity and maturity never surpassed by the thinkers of antiquity and the Middle Ages. His ideas had a decisive influence on this controversy.

2) The present study brings into focus a "school," that is, a specific intellectual milieu, because the savants who were involved in the debate—Robert Grosseteste (1168–1253), Roger Bacon (1214–94), and John Pecham (1230–92)—all produced their treatises on optics in much the same period, during the second half of the thirteenth century. Grosseteste's *De iride* dates to the 1240s, Roger Bacon's *Opus majus* and *De multiplicatione specierum* appeared in 1262–63, while John Pecham produced his *Tractatus de perspectiva* in 1269 and his *Perspectiva communis* in 1279. What is more, these three scholars were all affiliated with the Franciscan *studium* of Oxford. While there are no documents to show that Robert Grosseteste was a member of the Friars Minor, we know that he held a post as a lector with their

studium for a period of at least twelve years. Roger Bacon entered the order in 1251, after studying under Adam of Marsh, a close associate of Grosseteste who succeeded him as *magister regens* of the university at Oxford. The last of the three, John Pecham, became a Franciscan around 1248 with the encouragement of Adam of Marsh, under whom he had studied until 1257, the year in which he left the *studium oxoniense* for Paris (Lindberg 1972). Thus, there exists a filiation between Grosseteste, Bacon, and Pecham. Although historians do not all agree that Bacon was a direct disciple of Grosseteste (Bridges 1964 versus Lindberg 1983), they unanimously acknowledge the influence of the latter on the former; Bacon cites Grosseteste continually in his work, and often refers to him in the most laudatory of terms.[2] It has also been shown that Bacon and Pecham knew each other personally, both having boarded at the *studium parisiense* between 1257 and 1267.

These three scholars, each of whom would make a decisive contribution to medieval optics, belonged to the same intellectual milieu (Raynaud 1998), and a study of their divergent positions will expose not only the scientific content of the controversy at the time, but also the norms of rationality as a function of which this scientific content was chosen.

1. Extramission versus Intromission

The basis of the controversy over optics lay in two opposing theories regarding how vision works. Does the act of seeing involve the emission or the reception of visual rays by the eye? These two modalities were known by the terms *extramission* (light emanating from the eye) and *intromission* (light entering the eye). Herein lay, in a nutshell, the origin of the debate. To answer the question regarding the direction in which visual rays were propagated, our three Oxonian perspectivists (*perspectiva* was the Latin word for optics) could logically only choose one of the four solutions presented in the form of a Carroll's square (Figure 3):

Assuming that I represents "in" and O "out," there are only four possible combinations: (¬I, O) extramission alone; (I, ¬O) intromission alone; (I, O) a combination of intromission and extramission; and (¬I, ¬O) the negation of both theses. The fourth solution was not considered by the doctors at Oxford, although one can find an echo of the thesis "not I, not O" in the theory of the *medium* of light transmission proposed by Galen and adopted by his Arab successors, notably Ḥunayn Ibn Isḥāq, a pupil of Ibn Masawayh, who elaborated

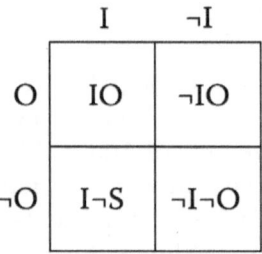

Figure 3. Diagram of the four possible modalities of vision.

on it in his *Book of the Ten Treatises on the Eye* (see Meyerhof 1928; Lindberg 1976, 38).

During the period in which we take up this controversy, none of these theses could be considered to be true, because there did not exist any experimental tests that would have been able to decide between them with certainty. Nor had any irrefutable logical argument been found that weighed in favor of one or the other. In a study devoted to aspects of this controversy in Arab thought, Lindberg (1978) demonstrated with perfect clarity that the choice between intromission and extramission was closely linked to fundamental questions concerning the very field of study to which one assigned the theory of vision. The determination of the direction of propagation of light rays in vision could, in this sense, depend on developments in the physics of optics, or in the anatomy and physiology of vision. Equally, other studies have shown that the preference for one thesis over the other had repercussions on questions of the psychology of perception. Our perspectivists nonetheless set out to justify their respective positions based on a more or less convincing argumentation (whose scientific tenability was, in addition, extremely variable, with arguments that oscillated between simple belief and demonstrable proof). Before turning to a consideration why each of our protagonists preferred one thesis over the others, it will be useful to review the armamentarium of arguments that they had at their disposition.

1. The Thesis of Extramission

The thesis of extramission, according to which visual rays were emitted by the eye, was the product of theories of vision that were first formulated in antiquity. The arguments advanced were the following:

1. The existence of phosphenes. The savants generally referred to an observation made by Aristotle, more or less in passing, with regard to the emission of light by the eye: "When the eye is pressed or moved, fire appears to flash from it" (*De sensu et sensato*, 437a). If the eye is capable of generating light even when the eyelid is closed, it was reasoned, the emission of this "visual fire" must be an intrinsic property of the organ of sight. And there seemed to be no reason why it should be otherwise under the normal conditions of vision, that is, when the eyelid is open.

2. *The selectivity of the eye.* The supporters of the thesis of extramission in the Middle Ages drew massively on another argument in the antique texts, one of the earliest occurrences of which may be found in a commentary on Euclid's *Optics* by Theon of Alexandria: "If the sense of seeing is made of the images that continuously flow from all bodies, that our sense is suffering, what is the reason why he, who seeks a needle or looks through a book, does not notice the needle and all the letters?"[3] Taking the example of the book, if one's gaze is fixed on a single word, only those immediately adjacent to it can also be read; the other words on the same page will not be deciphered. This argument, which apparently weighs in favor of extramission, is in fact invalid because it confounds three distinct processes: the propagation of rays (which is a question of physical optics), the gaze (i.e., aiming the eyes at an object, which involves a coordinated motion of the eyes), and focus (i.e., producing a clear image, which is due to the action of the ciliary muscles on the capsule of the crystalline lens). These three processes would not be individuated until the discoveries in physiological optics made in the nineteenth century.

3. *The spherical form of the eye.* This argument, which rests on a medieval philosophical conception of the correspondence between the form of an organ and its function, had already been formulated by Damianus: "That it is thanks to a diffusion of light issuing from us which strikes the objects of sight, is demonstrated to us by the form of the eye. This form is neither concave, nor disposed for the reception of some thing [. . .], but it is spherical" (in Mugler 1964, 322). Here the thesis of extramission is supported by the paralogism of the correspondence of forms: the convex shape of the eye is not adapted to "receive" externally generated visual rays.

4. *The phosphorescence of the eye of the feline.* The fact that cats can see in the dark has often been used as an argument in favor of extramission. Savants often might cite a thesis even when they were aware of the existence of objections to it. Like a number of his predecessors,

Pecham addressed the question of the phosphorescence of the cat's eye. He writes: "However, the cat is said to have so much light [in the eyes] that is sufficient to correctly represent the objects, and I think this to be true."[4] In his second treatise Pecham (1279) would add—this time making a concession to popular belief: "Sight occurs only through a transparent medium [...] No density completely prohibits the propagation of powers and [visual] species, although that propagation may be imperceptible to us. Hence lynxes are said to see through walls."[5]

A comparable notion can be detected in Bacon's *Opus majus* (V, I, V, 1). If vision in cats can dispense with a transparent medium of propagation, it is because their sight is "piercing" and therefore active. This confusion between modalities of seeing can be attributed to the fact that cats possess a unique anatomical feature—the *tapetum lucidum*, a membrane located on the choroid which reflects the light received in the eye. Therefore, the argument cannot be extrapolated to man.

5. *The corruption of mirrors.* The popular belief that mirrors could reflect subliminal states has existed since antiquity and even Aristotle—despite his support for the thesis of intromission in *De sensu et sensato*—recounts an example: "At the same time it becomes plain from [the mirrors] that as the eye is affected [by the object seen], so also it produces a certain effect upon it. If a woman chances during her menstrual period to look into a highly polished mirror, the surface of it will grow cloudy with a blood-coloured haze" (*De insomniis*, 459b). In the context of medieval philosophy, this superstition was used to support the thesis of extramission: visual rays were capable of transmitting the sanguine nature of the eye to the diaphanous medium and to external objects such as mirrors. This pseudoproof is still cited by John Pecham (*Tractatus de perspectiva*, IV, 7).

6. *The recessing of the eye.* This justification as well can be found in Aristotle: "But the cause of sharpness of vision in seeing objects at a distance is the location of the eyes, for the prominent eye does not see well at a distance. And the deep eye sees remotely, for its movement is neither divided nor consumed, and the visual power comes out of it and goes straight to the things seen."[6] At first glance, it is difficult for us to grasp the relationship between the recessing of the eyes within their orbits and the thesis of extramission. This link was nevertheless drawn at the time. Let us assume that the seat of the faculty of sight is situated in the chiasma. The shorter the distance between this point

and the eye, the lower the loss of visual power along the optic nerve, and the better one will see. This is why the problems with vision experienced by those suffering from exophthalmos could be used as proof of the thesis of extramission. Notably, this idea was accepted, or at least reported, by Robert Grosseteste (*De iride*, 73), Roger Bacon (*Opus majus*, V, II, I, 1), John Pecham (*Tractatus de perspectiva*, IV, 74), and Roger Marston (*Quodlibeta quatuor*, 58).

As can be observed, all of these arguments verge on paralogism or popular belief. But in order to understand and assign due weight to the terms of this debate, we must refrain from applying the same criteria that would be applied today. Let us accept them for the moment as justifications for the thesis of extramission.

2. The Thesis of Intromission

The very fact that this controversy arose is proof that neither of the theses managed to win an immediate consensus, even among scholars from the same intellectual milieu. For centuries they were regarded as opposing theses but of equivalent weight, each one corroborated by an ensemble of reasons and pseudoreasons. The medieval supporters of the thesis of intromission, according to which the eye functioned by capturing the visual rays emitted by the external world, opposed the thesis of extramission by presenting a new set of arguments:

1'. *The absence of night vision in man.* This argument was known to students of optics in various forms, but most of them were based on different readings of Aristotle. In fact the philosopher from Stagira rejected the position of Plato in favor of the thesis of visual fire on the following grounds: If vision were the result of light issuing from the eye as from a lantern, why should the eye not have had the power of seeing even in the dark?" (*De sensu et sensato*, II, 437b). If the human eye is not capable of seeing at night, then it cannot be compared to the eye of the cat and therefore the extramissionist model of vision, at least on these grounds, is false. Indeed, this has always been the main objection against the thesis.

2'. *The pain caused by bright light.* Here the learned doctors of the Middle Ages were relying on the authority of Alhacen, whose works had recently been translated into Latin, and who refuted the thesis of the emission of visual rays. The Arab savant writes: "We find that when our sight fixes upon very strong light-sources it will suffer intense pain and impairment from them, for when an observer looks at the body of the sun, he cannot do so properly because his vision will suffer from its light."[7] If it is true that the eye emits a visual fire, why should it suffer more when focusing on the sun than on a green meadow? The different in affection (tears in the eyes, temporary blindness due to the glaring light, etc.) indicates that the eye does not emit light, but receives it in greater or lesser quantities in a trajectory that travels from the visible object to the eye.

3'. The absence of instantaneous propagation of light. The question as to whether the propagation of visual rays occurs instantaneously or over an interval of time, however brief, received different answers in different periods. In the thirteenth century, the mention of this point was accompanied more often than not by a response that presupposed a propagation time. As Bacon writes: "Therefore, the remaining possibility is that light is multiplied in time, and likewise every species of a visible thing and of the eye."[8] Pecham repeated this same argument in a slightly different form.[9] In his variant he pondered: If the visual faculty works by means of rays emitted by the eye, how is it that we can see simultaneously an object that is very close and one that is far away? When you open your eyes, it would be logical for the world to reveal itself gradually in a sequence of concentric layers, from close up to far away. In fact, however, nothing of the kind occurs and therefore one must conclude that the eye does not emit visual rays.

4'. The impossibility of immensely long visual rays. This argument, which was already known to the savants of antiquity, was cited in the Middle Ages by advocates of the theory of intromission. John Pecham, for example, developed it to oppose the theory that visual rays were the propagating vector of the images of visible bodies. He writes: "Therefore the visible object need not to be sought out by rays as by messengers. Moreover, how would any power of the eye be extended all the way to the stars?"[10] In effect, if one grants the notion of extramission, the eye is not powerful enough to send a visual influx to the sphere of fixed stars, whose distance was estimated by Roger Bacon to be 130 million kilometers.[11] And even if it could, this hypothesis would fall before the preceding argument.

If one draws up a list of all the arguments that were conceived in support of the intromission and extramission theories of visual rays, both qualitative and quantitative differences emerge. Quantitative differences in that we can identify at least six arguments in favor of the thesis of extramission and only four supporting the opposing thesis. Qualitative differences as well, because manifestly the various arguments are not of equivalent weight, a point that we will address further on.

2. The Arguments Presented by the Three Oxonians

Robert Grosseteste, Roger Bacon, and John Pecham took part in the controversy that was revived in the thirteenth century between two contrasting theories of vision, one claiming that the eye emitted rays, and the other asserting with equal conviction that the eye was a passive recipient of the rays that were the source of vision. All three savants were well acquainted with the stock of arguments furnished by the sources reviewed above, many of which they cite explicitly and others of which emerge from a comparative analysis of the texts (Raynaud 1998a; 2014). At this point it is important to examine the conclusions that each one deduced from his sources. Confronted with the same identical problem, the three Oxonians imagined very different solutions.

1. The Position of Grosseteste

Robert Grosseteste viewed the controversy in these terms: "Therefore, the natural philosophers, who touch the nature of sight on one side as a passive [process], say that vision is effected by intromission. The mathematicians and physicists, however, who consider the nature of sight on the other side, say that vision is effected by extramission." Basing himself on Aristotle's *De animalibus*, Grosseteste decides in favor of the thesis of pure extramission: "The true perspective, then, poses the emission of rays."[12] It should be noted that this position was the result of a partial reading of Aristotle. Careful comparison of his treatises shows that in reality Aristotle oscillated continually between the hypotheses of extramission and intromission. In the treatises *De insomniis* and *De animalibus* he argued in favor of extramission, whereas he expressed himself more clearly in favor of the second thesis in *De sensu et sensato*. Elsewhere he concludes: "Rather, the color sets a transparent thing, such as air, in motion, and by this, the sense organ is moved" (*De anima*, II, 419a). Grosseteste ignored these inconsistencies and adhered to the accepted, homogeneous reading that was made of the Aristotelian texts before the dissemination of critical editions prepared by the Arabs. Grosseteste cites Averroes (Ibn Rushd), Avicenna (Ibn Sīnā) and Albumazar (Abū Ma'shar), but he did not have access to Alhacen (Ibn al-Haytham) and therefore never refers to his theory.

2. The Position of Bacon

Contrary to his predecessor, Roger Bacon was well versed in both the texts of the Greco-Roman tradition and the treatise of Alhacen. Therefore, unlike Grosseteste he could not argue for one thesis to the exclusion of the other. Since his authoritative sources disagreed, he refrained from accepting one thesis and completely rejecting the other, instead proposing a conciliatory synthesis of the two. As he observed, some believe that the eye launches rays while others contest this opinion; as a consequence, he suggested that images cannot penetrate the eye unless they are "aided and called upon by the visual faculty [. . .]." This hybrid solution combined intromission and extramission and Bacon defended his thesis by explaining that there was no inherent contradiction in both being true simultaneously: "There is no confusion of those species, nor mixing, nor do they unite, for they are not of the same genus or species."[13]

3. The Position of Pecham

In contrast, John Pecham, who was well acquainted with the texts of Alhacen—through Bacon with certainty and probably also through the works of Witelo (Lindberg 1971, 81)—adopted a clear position in support of the thesis of intromission. Taking up one of the arguments of Alhacen, he writes: "The action of the visible object on the eye is painful. It is proved as follows. Because the action of the visible object on the eye is of a single kind and the action of bright lights on the eye is sensibly painful and injurious, it follows that all actions of light are painful and injurious, even if they do not seem to be."[14] If visible bodies emit their own light, then the extramission of rays from the eyes is not necessary. On these grounds he attacked all of those who argued that emission was necessary to the act of seeing: "Mathematicians exert themselves unnecessarily by assuming that sight occurs through rays issuing from the eye [. . .] Therefore it is superfluous to posit such rays."[15] It may be noted that Pecham arrived at this conclusion based not simply on his reading of Alhacen's De aspectibus, but after careful comparison with the texts of the Platonicians, Saint Augustine, and al-Kindī, who all sustained a thesis that was contrary to that of Alhacen. Pecham recognized therefore that the emission of rays by the eye was useless to the act of seeing (without, all the same, denying that the eye could occasionally emit such rays). This lingering residue of the thesis of extramission can be explained by the phenomenon of phosphenes (see argument 1), whose existence he does not deny and in which one unmistakably finds the emission of light by the eye that does not contribute to the act of seeing. The Oxonian concludes: "It is evident, therefore, that there is some kind of emission of rays, but not of the Platonic type such that rays emitted by the eye are, as it were, immersed in the visible form and then returned to the eye as messengers."[16] He regarded extramission as an accidental phenomenon with no specific functionality.

3. Sociohistorical Analysis

As we have seen, Grosseteste, Bacon, and Pecham adjudicated differently on the direction of visual rays, and this: despite the fact that they produced their treatises within a short time of one another; despite their frequentation of the same *studium oxoniense*; despite their shared vows of obedience to the same religious order; and despite the fact that they were addressing, with comparable intellectual abilities, the same scientific problem. Grosseteste embraced the thesis of extramission; Bacon formulated a conciliatory synthesis; and Pecham defended the thesis of intromission. But this entire set of facts controverts, to say the least, the SSK hypothesis that the contents of scientific statements are socially determined, as emerges clearly when one examines the prevailing norms of rationality and the scientific content that would have been considered receivable in the period when this debate took place and in the circles to which these savants belonged.

1. The Medieval Norms of Rationality

The controversy being described did not unfold in just any scientific context. Today we acknowledge a scientific statement to be true either because it is logically consistent or because it corresponds to reality. In the Middle Ages, however, the word "truth" had not two, but three distinct acceptations. This threefold definition was enunciated by Roger Bacon in *Compendium studii philosophiae*, where he asserts: "*Licet per tria scimus videlicet per auctoritatem et rationem et experimentum* [We may know by three ways, that is to say, by authority, reason and experiment]." Even if, as an advocate of experimental science, Bacon declared his preference for knowledge based on the third modality, he admitted the concurrent existence of all three forms of knowing.

The modality *per rationem* is equivalent to what we would today call the validity of reasoning or logical consistency. To display rationality signifies that one is making use of accepted forms of reasoning. The syllogistic inference, like the rules of *modus ponendo ponens* and *modus tollendo ponens*, belong to this class. This quality was an absolute requisite for the Oxonians, in particular Bacon, who declared mathematics to be the canon of logical reasoning for the natural sciences.[17] However, because these logical forms were in no way determined by the contents of the statements, they could lead to false results if applied to erroneous premises (*Opus majus*, IV, I, 3) as Popper, an advocate of the theory of correspondence, does not fail to point out: "Now it is quite obvious that the truth of a theory cannot possibly be inferred from its consistency" (Popper 1985, 264). This is what justifies the second medieval conception of truth.

Per experimentum is equivalent to the modern notion of experimental proof. A statement is true when it corresponds to reality and to the empirical facts, in short when it is objectively based. The disquotational and correspondence theory of truth by Tarski (1933) offers a contemporary equivalent.[18] An approximate form of this definition was admitted as a norm for the construction of statements in the Middle Ages. It is encapsulated in the formula of Thomas Aquinas: *adaequatio rei et intellectu.* Bacon as well accorded great importance to direct experience, which he sometimes placed on an equal footing with reasoning, and sometimes even before it (*Opus majus* VI: *De scientia experimentalis*).[19] He writes in *Compendium studii philosophiae: Necesse est per rerum ipsarum experientias certificari veritatem*, that is, "Truth must be certified by experiments on the things themselves." Bacon did not rely on experimentation as systematically as he claimed, but it did seem to him to be the most essential of the three concurrent forms of truth.

Finally, *per auctoritatem* corresponds to our notion of arguments from authority. Medieval scientific texts show that in the period when the debate over optics was revived, the word "truth" could also be used to qualify a statement of metaphysical or religious knowledge, thus a truth that evaded the domain of experimentation. The medieval notion of authority was inseparable from this acceptation, the authority in question being the Holy Scriptures—qualified as the "Primary Truth" of which God Himself was supposed to be the author (Thomas Aquinas, *Summa theologica*, quaestio 1, 10)—followed by the patristic literature and, by extension, the works of the classical authors. This transposition does not always mean that there was a confusion in the scholars' minds between the unconditional truth of the Scriptures and the conditional authority of the philosophers (*Summa theologica*, quaestio 1, 8). But from this, as Leclerc has perceptively observes, "The term *auctoritas* could be applied just as well to the intellectual weight of the classics in profane culture as to the Scriptures in sacred knowledge. The authority was the text [. . .], worthy of belief, the mandatory path toward the truth" (1996, 99). Such is how the famous argument *per auctoritatem* could be understood, before it was rejected as a norm of scientific rationality. This particular form of truth, unlike the other two, may be compared with the notion of traditional rationality (Boudon 1990), because the uncontested value of the texts of Aristotle and Saint Augustine flowed from the medieval scholastic tradition of the study and discussion of texts.

As a consequence, in their treatises on optics Grosseteste, Bacon, and Pecham would all have felt obliged to certify their statements taking into account these three principles of truth. In the thirteenth century, a statement that was supported neither by logic nor by experimental evidence nor by an authoritative author was considered to be an irrational statement. This position was accepted by all three Oxonians, as can be seen if one turns to the main causes of error identified by Roger Bacon before he formulated his theory of optics (*Opus majus*, I, 1).

2. The Authority of Saint Augustine

Since the arguments to be found in the definitive sources were regularly used in the construction of scientific statements in the Middle Ages, it would be pertinent to examine which authorities were perceived as having something to contribute to the discussion on the direction of visual rays and on what grounds they could be incorporated into the debate.

The religious order to which all three of our Oxonian scholars belonged provides the first indication. Bacon and Pecham were Franciscans, while Grosseteste was closely associated with the order, having served as a lector for twelve years (1223–35). There is ample evidence to show that a large number of Franciscan scholars adhered to the school of philosophical-religious thought referred to as Augustinism, which was based on the doctrine of Saint Augustine combined with other compatible sources, most notably the works of medieval Arab thinkers. Traditional Augustinism implied the corollary rejection of Aristotelianism.

Many indices confirm the Franciscan order's strong links to the doctrines of Augustinism.

Primo, in a period when Augustinism was held in the highest regard in the intellectual circles of the Holy See, a series of Friars Minor were nominated as lectors at the Sacred College, (i.e., the papal university). John Pecham (1277–79), his pupil Matteo of Acquasparta (1279–87), and then Pietro de Falgario (1287–91), Giovanni da Morrovalle (1291–96), Gentile da Montefiore (1296–1300), and William of Gainsborough (1300–02) all served successively as lectors at the Sacred College (see Raynaud 1998).

Secundo, in a study of the opposition between traditional Augustinism and Aristotelianism, Callebaut (1925) examines the connection between philosophical doctrine and the decisions taken by the ruling ecclesiastical institutions, and highlights the role played by John Pecham in these philosophical-religious discussions. Pecham's allegiance to Augustinism was clear; his *Tractatus de perspectiva* opens with the words *Dixit Augustinus*, and his argumentation is supported throughout by extensive quotations from the Psalms, the Book of Wisdom and Ecclesiastes.

Tertio, while less marked, signs of Augustinism can also be detected in the works of Grosseteste (who cites Augustine's *De trinitate, sermo CLV*, etc.) and Bacon (who refers to *De doctrina christiana, De musica, Soliloquia*, etc.).

At the very least, these clues would seem to confirm a doctrinal consensus among the perspectivists at the school of Oxford.

The defenders of Augustinism—who therefore included Grosseteste, Bacon, and Pecham—had some reason to cite the authority of Saint Augustine on the question of extramission, because the theologian expressed his own views on this point clearly in many of his works (*De genesi ad litteram*, XII, 32; *De trinitate*, XI, IV, 4; *De musica*, VI, V, 10). For example, in *De genesi* we find: "Light is initially diffused through the eyes, then it shines forth in rays from the eyes to gaze upon visible objects."[20] This is in no way to be regarded as a random statement appearing by chance, in a discourse on philosophical doctrine. The phenomenon of extramission was deduced on theological grounds.

In his view, vision was an active faculty and complete in itself. It had no need of an external, causal influx such as light, because vision emanated directly from God through the intervention of the soul. The thesis of intromission posed a threat in that it might be interpreted as a negation of the power of the divine. It could therefore, if not destroy, at least undermine the central doctrine of Augustinism, which was referred to as "emanatism." *First contradiction: if the possibility existed for Grosseteste, Bacon, and Pecham, who were all affiliated with the Franciscan order, to establish the direction of visual rays on theological grounds by invoking the authority of Saint Augustine, why did they not do so?*

Contrary to what a number of SSK studies have concluded, the political-religious context in this controversy does not weigh in favor of their central thesis of social determination. It runs *counter* to the explanation generally provided by the anti-Mertonian school regarding how controversies are resolved, as for example in their analyses of the dispute over spontaneous generation (Farley and Geison 1974; Mendelsohn 1987; Latour 1989; Lécuyer 1990; Collins and Pinch 1993; and Vinck 1995).

It is possible to distinguish another nuance in the medieval reception our three Oxonians' theses. If one accepts the SSK assumption that the success of a theory can be explained by the influence of politico-religious factors, it must then be presumed that within the scientific milieu there are individuals or groups whose interests will provide a conduit for these extrascientific biases. In the controversy between the homogenists and the heterogenists, the supposed predisposition toward Pasteur of the Academy of Sciences' expert commission would represent an example of the transmission of such interests. And yet, even though social interests tied to religious ideologies may have been present in the controversy between intromission and extramission, these interests certainly did not influence the debate as heavily—and certainly not in the same direction—as in the controversy over spontaneous generation discussed in chapter 2, or in the controversy on phrenology in which Shapin (1979) sought to demonstrate the influence of political interests. The model drawn based on the existence of these interests fits awkwardly with the way in which the medieval controversy over optics unfolded. To be convinced of this, it is sufficient to conduct a detailed comparison of the respective situations of Bacon and Pecham.

Bacon sought to resolve the matter by proposing a synthesis of the opposing theses, but sometimes he went so far as to adjust the wording in his citations from the texts of Saint Augustine in order to eliminate ambiguities or add emphasis to pronouncements that could be linked to the thesis of extramission. Notably, he writes, "In *On Music*, book 6, Augustine asserts that the species of the eye is engendered in air as far as the [visible] object,"[21] a proposition that is not expressed in such unambiguous terms in the original. At such moments Bacon was writing as a defender of Augustinism, a perfectly understandable position on his part.

What is less comprehensible, and hence quite interesting, is the disjunction between Bacon's choice of a synthesis of an Augustinian hue and the reaction of the groups in his immediate circle or "core set." Bacon's ideas on optics were not in fact so very heterodox and should have earned him the protection of his religious order. But this was not the case. Bacon was regarded with suspicion by his superiors, in particular by the order's general minister, Girolamo d'Ascoli, to the point that he was placed under surveillance for ten years at the *studium parisiense* (1257–67) for reasons that were never made clear; we find only the vague accusation that he was guilty of "some suspect novelties" *(quasdam novitates suspectas)* (Lindberg 1983, xxv).

John Pecham received quite a different treatment. Despite his sometimes blunt championing of the thesis of intromission—including, for example, direct criticism of the thesis of extramission that included some pointed invectives aimed at his opponents—he does not seem to have been penalized by the Franciscan order. In fact, his ascent was rapid in what proved to be a brilliant ecclesiastical career. Received into the faculty at Paris as a *master regens* in 1269, he returned to the *studium oxoniense* in 1271–72 where he served as the eleventh *master* of the order. He was appointed Provincial Minister of England in 1275, before being entrusted with the position of lector at the Sacred College in 1277. *Second contradiction: Bacon, who was open to the doctrines of Augustinism, was placed under surveillance whereas Pecham, who championed theories of medieval Arab scholars that invalidated Saint Augustine's doctrine of emanatism, was protected and promoted by his religious order.*

Here again the results contradict the conclusions of SSK studies on scientific controversies. If the authorities of the religious order in Oxford had acted in conformity with their own interests, which included their preference for the doctrines of Saint Augustine, they would have supported Bacon and sequestered Pecham instead.

3. Questions of Authority

The ensemble of factors that are undeniably connected with this controversy—the reliance in medieval science on arguments drawn from authoritative sources; the adherence to Augustinian doctrine by the Franciscans; and the support of Augustinism for the thesis of intromission—would appear to offer an ideal canvas for relativists seeking to illustrate the influence of social factors, or a combination of factors, on scientific theories. Yet these facts fail to explain: (1) why Grosseteste, Bacon, and Pecham should have disagreed, because if they belonged to the same intellectual milieu one would have expected them to embrace the same theory; (2) why none of the three invoked the authority of Saint Augustine to settle the debate, when arguments drawn from authoritative sources were considered to be decisive in the construction of scientific statements and the texts of Saint Augustine certainly appeared pertinent to the question; (3) why was there a negative rather than a positive correlation between the positions held by Bacon and Pecham on Augustinian doctrine and their career paths within the religious order?

These difficulties all point to an unexpected conclusion: our three authors envisaged that the question of the direction of propagation of the rays could be resolved *rationally*. Still, this begs a question that must be faced before we continue: Could the adoption of the thesis of intromission be interpreted simply as the substitution of the authority of Alhacen for that of Saint Augustine? If so, Bacon and Pecham would not have been defending the thesis *per rationem et experimentum* but *per auctoritatem*, reflecting the consensus that was forming in this period on the work of Arab authors, and in accordance with rules of doctrinal preference that were absolutely identical to those that would have conditioned the adherence of the Oxonians to Saint Augustine. The only fact that might support this notion, however, is the existence of a certain parallelism between the diffusion of Arab optics in the West and the gradual increase in the number of adherents to the thesis of intromission. Grosseteste refused the thesis in 1240, Bacon half admitted its veracity in 1262, and Pecham defended the thesis in its entirety in 1279, in a progression that reflects the increasing penetration in Western Europe of Alhacen's *De aspectibus*. Grosseteste does not cite Alhacen even once, Bacon quotes passages from Alhacen in various parts of his treatise, and Pecham had a thorough knowledge of *De aspectibus* from the work of Bacon and Witelo, and probably also directly through the translations of William of Moerbeke, confessor and chaplain of Pope Clement IV, who was living in Viterbo in this period (Paravicini Bagliani 1975).

Insofar as the learned doctors of Oxford were allowed to study and discuss the works of Alhacen, their position became equated with that of the Arab savant. But was this a debate in which allegiance simply shifted from one authority to another, or was it a critical discussion that led to the acceptance of the rational arguments presented in *De aspectibus* which, they realized, were strong enough to overturn the opinions of Saint Augustine? A study of the context and tenor of the Latin translations from the Arab show that these versions were in no way intended to overthrow the authority of Christian thinkers. In fact, quite the contrary may be argued. (1) The majority of the translations were made by and for the ecclesiastic institutions, beginning with the work of William of Moerbeke who was an intimate at the papal court. (2) If the authority of Alhacen did simply substitute for that of Saint Augustine, we should observe a decrease in the number of citations of the Christian theologian in the last of the series Grosseteste—Bacon—Pecham. Instead, we find that even Pecham quoted from Saint Augustine frequently and extensively. (3) If the thesis of intromission had been the object of a discussion *per auctoritatem*, the texts of the Oxonians would not have focused to such a degree on arguments based on geometry and experiments in optics. This last point therefore can be dismissed as well.

We may first of all take the measure of the logical and geometric arguments presented in the treatises of Bacon and Pecham. In the vast panorama of styles of scientific thinking, Crombie (1994, 313–423) concentrated on the development of medieval logic beginning with the commentaries on Aristotle. He noted that the form of the proof by contradiction (*reductio ad absurdum*) was often employed at Oxford. In addition to the categories of logic, the distinctions between certainty and conjectural habit (*certitudo, conjecturativus habitus)*, intention and claim (*inductio, postulatio*), and the concepts of assumption, inference, proof (*suppositio, argumentum, probatio*) were of common usage. The role of experimentation is also documented. Bacon had special mirrors made for his optical experiments and even expressed some disquiet as to the cost of the instruments that he had ordered.[22] Following Alhacen's lead, Pecham conducted studies using the *camera obscura*, contrived devices for the projection of light (*Perspectiva communis*, I, 5–7), and repeated the Arab scientist's experiments on refraction. Crombie (1994, 313–423) also provides details on the terminology specific to experimental knowledge (*cognitio experimentalis*) that came into use in the Middle Ages: *inductio, modus procedendi, probatur, falsificatur*, etc.

It is of interest to note here that Crombie drew a large number of his examples from the treatises on optics of Alhacen and his Latin commentators, Bacon and Pecham.

It will be admitted that the mere repetition of an argument drawn from an authoritative source does not require that a scientist provide much in the way of logical or experimental support for a thesis. It suffices to state the thesis and refer to this authority. The absence of such argumentation allows us to consider the intromissionists' response to their opponents' thesis as the result of a rational process of reasoning.

4. A Rational Choice?

As we have already noted above, the arguments presented by the partisans of extramission and intromission were marked by fundamental differences that were both quantitative and qualitative in nature. Notably, a comparison shows that certain of the extramissionist arguments could be easily refuted at the time. This is the case with argument 1 (the existence of phosphenes) and argument 4 (the retroreflection of the cat's eye), which are directly contradicted by argument 1′ of the intromissionists (the absence of night vision in man). The mode of vision in cats cannot be transposed to man, while the phenomenon of phosphenes does not prove that vision is due to pure extramission, only that an "internal fire" could "collaborate" in visual perception. Moreover, while the assertions contained in arguments 1′ 2′, and 4′ are not opposable, arguments 3, 5, and 6 could—without rejecting them outright—be judged doubtful. In fact, argument 3 could be reinterpreted in favor of the intromissionist thesis, if one proposes that sphericity is a necessary condition for the movement of the eye within its orbit rather than a condition for the emission of visual rays. Argument 6 could be refuted on the basis of the observation of individuals with bulging eyes who nevertheless have normal vision. As a consequence—without in any way prejudging the adherence to these ideas—it may be observed that the arguments for extramissionism, while more numerous, could be more easily challenged and this might explain the ultimate success of the thesis of intromission.

There remains the problem that in the thirteenth century the thesis of the intromission of luminous rays was derived from a train of reasoning that could be partially contested. Argument 3′—the propagation of light over time—presents a residual difficulty. By linking a proposition of physiological optics (How does one see?) to a proposition of physical optics (How does light propagate?), it assumed the form of a proof by contradiction (*reductio ad absurdum*). The thesis of extramission posits that, when one opens one's eyes, the visual ray must make its way toward the visible object before returning to the eye. Under these conditions, the eye should perceive objects that are closer *before* objects that are further away, because the ray has a longer distance to cross and should take more time to travel before it returns into the eye. As a consequence, one would expect objects to reveal themselves to our sight in "concentric layers"—a tree should appear before the clouds behind it, clouds before the stars, and so on. However, experience proves the contrary: when one opens one's eyes the tree, clouds, and stars all appear simultaneously. The thesis of extramission is therefore refuted (*refellitur*).

This argumentation would appear to be implacable, but it is actually supported by an "additional hypothesis" relative to the propagation of light. Most students of optics in the Middle Ages assumed, following the lead of Plato, Hero of Alexandria, Ibn al-Haytham, al-Bīrūnī and al-Fārisī, that light has a finite velocity, and that its speed could be affected by the density of the medium through which it was passing. Notably, this position was held by both Roger Bacon and John Pecham at Oxford. In his *Opus majus (pars V: Perspectiva)*, Bacon writes: "Therefore, if the multiplication of light is instantaneous and does not occur in time, there will be an instant without time, since there is no time without motion. But it is impossible for an instant to exist without time, just as it is for a point to exist without a line. Therefore, the remaining possibility is that light is multiplied in time, and likewise every species of a visible thing and of the eye. Yet this time is not sensible or perceptible by sight, but insensible [. . .]."[23] Bacon used a geometric argument to justify the notion of the temporal propagation of light, in which there transpires an atomistic theory of space and time—just as a point is in reality an infinitely small segment, an instant will always be an infinitely short interval of time. John Pecham granted this thesis as well, even if he did not justify his position with a geometric proof.[24] On the contrary, other perspectivists assumed—in line with the Aristotelian model of "asomatic light" (*De sensu et sensato*, 447a)—that light rays propagate with infinite speed. In his commentary on *Fiat lux*, Basil of Caesarea broaches this point in passing: "For [light] reaches up to the sky and up to the heavens, and illuminates at large all parts of the world in a short space of time [. . .] Such is its very nature, that is, thin and transparent, so that the light passing through the air requires no delay."[25] In the late sixteenth century, authors including William

Gilbert (*De magnete* VII) and Caspar Schott (*Magia universalis*, I, 72) endorsed the notion of the instantaneous propagation of light. With this additional hypothesis, the initial contradiction in the thesis of extramission was nullified *tout court*.

At this point two observations must be made.

Primo, it was impossible within the medieval thought system to resolve the question as to whether the propagation of light is instantaneous or occurs over time. The first experiment to actually measure the speed of light was conceived by Galileo in 1638 (*Discourses and Mathematical Demonstrations Relating to Two New Sciences*, I, 87–88). But his experiment was somewhat rudimentary[26] and did not allow the scientist to decide one way or the other on the question of the propagation of light. Indeed, the first reliable measurement of the speed of light was made by Ole Rømer, who timed the eclipse of Jupiter's satellites in 1676. Finally, in the nineteenth century measurements based on astronomical observations were abandoned for more convenient methods using terrestrial instruments to study the periodic darkening of a beam of light, such as Fizeau's rotating toothed wheel (1849) and Foucault's rotating mirror (1850). These experimental procedures were not available to medieval scholars.

Secundo, if the speed of light was infinite rather than finite, *this proof would have refuted in equal parts the thesis of intromission and that of extramission* in favor of a third thesis that had already been put forward by Ḥunayn Ibn Isḥāq (denoted ¬I, ¬O). In fact, the notion of instantaneous propagation signified that the visual ray would occupy the two extremities of the eye and the visible object *simultaneously*. There would therefore be no movement in either direction. The argument of the instantaneous propagation of light could not be used by the extramissionists, because it would have completely demolished their thesis.

Therefore, in the thirteenth-century debate over optics, arguments vitiated by uncertainties existed on both sides, but *this uncertainty was asymmetrical* because some of the arguments for extramission could be salvaged and reused by the intromissionists, whereas the contrary was impossible. Finally, the superiority of the thesis of the reception of light was linked to a unique additional hypothesis that was entirely reasonable, if not experimentally demonstrable.

The settlement of this medieval controversy was thus constructed using the asymmetric stock of arguments accumulated by the two sides, which was itself conditioned by the texts available to the protagonists. If the thesis of intromission gained credibility in the thirteenth century, it was thanks to the asymmetry in the argumentation caused by the diffusion of Arab optics, and in particular Alhacen's *De aspectibus*. From a formal point of view, this asymmetry demanded *the replacement of one synthetic proposition for another, more likely synthetic proposition*.[27]

If one considers the question of the adherence of medieval thinkers to these theories, what we have here is a typical case of belief revision (Gärdenfors 1992; Walliser and Zwirm 1997). According to the typology presented by the SSK school of thought, the rejection of the extramissionist thesis would fall into the category of a change in belief through revision.[28] Following the work of Radnitzky (1987), Raymond Boudon (1990, 149) examined several cases in which a scientific question was decided long before any decisive proof of the solution was available. This type of settlement tends to occur in situations where we find a *closed* ensemble of theories (rather than an *open* set of theories, as assumed in the orthodox interpretation of Popper).[29] The case of two mutually exclusive theories is the most typical example, to the extent that one can then apply the rule of disjunctive deduction (see below, in the Conclusion). The resolution of the debate between extramissionists and intromissionists can be considered to be of this type. The models of belief revision suggest that the rejection of the Augustinian thesis can be explained as a rational choice that did not rely on any external determination.

4. Conclusions

Inversely to what would be predicted by a sociological reconstruction based on relativism, the intromissionists arrived at their positions by a rational examination of the problem and *against* the influence of extrascientific factors. This is why, at the conclusion of our analysis, we find a clear disagreement between the principles of relativism and the propositions demonstrated above: the absence of a consensus between the three Oxonians; the ease with which the Augustinian argument was relinquished during the course of the debate; and the fact that a defender of the theory of intromission could advance to the highest positions in the order.

To resolve these contradictions, it is necessary to renounce the tenets of the Strong Programme (Bloor 1976), the Empirical Programme of Relativism (Collins 1981), and the Actant-Network Theory (Callon and Latour 1991). We have shown that the social determination of scientific content, the influence of social structures on the procedures for limiting the extent of a disagreement, and the mobilization of more or less extensive networks in order to defend a theory are all far from being generalizable assumptions. The medieval debate over the direction of propagation of light rays unfolded in a completely different manner. It presents a striking particularity in that *knowledge in this case should have been socially determined or influenced by theological factors, but in fact it was not.* The scholars engaged in the debate over optics placed their faith in logic and experimentation rather than in the assertions of authoritative figures. This result calls into question the attempts by relativists to demonstrate at any cost the influence of rhetorical, ideological, or religious factors on the pursuit of knowledge *within the context of institutionalized science,* where the arguments of authority are no longer automatically accepted as common currency. This outcome also raises once again the question of the historicity versus the ahistoricity of the norms of rationality.

1) An opposition has been created between the relativists and the rationalists, the former maintaining that scientific statements should, and the latter insisting that they should not, be subjected to sociological analysis. Our study of the medieval controversy between the intromissionists and the extramissionists demonstrates the weaknesses in the relativist-constructivist approach to knowledge. One must always bear in mind that "scientific content" is never homogenous, because it can be divided at the very outset into factual versus theoretical statements (Lakatos 1980). Equally, one can distinguish between testable versus nontestable statements. Testable statements are those that can be judged logically or empirically. If I say: "The deep-set eye has a piercing gaze," *experience* will show me that there are both sighted persons and blind persons with bulging eyes, just as there are sighted persons and blind persons with deep-set eyes. Should I declare, "If the eye emits a visual flux, I ought to see closer objects before more distant objects when I open my eyes," I can demonstrate that this proposition is *logically* true. Such statements are testable, and when they are tested they can be shown to be either true or false (arising out of errors of observation or faulty logic). If they are false, sociology might eventually have something pertinent to say regarding the factors that were implicated in the construction of the error.

As for nontestable statements, they can impact in two places on a theory; on the *axioms* and on the *assumptions,* which are not discussed by the scientific community as a consequence of accepted conventions or norms. Therefore, in the reasoning process *examples* or *ad hoc hypotheses* become determinant. Non-testable statements cannot be subjected in the same fashion to sociological analysis, because they only "marginally" form part of a scientific theory. Nontestable statements, whose value in terms of truth cannot be established, are sensu stricto "ascientific statements." On this point, sociology could eventually have some observations to contribute regarding the interests or beliefs that led to such statements being adopted. When I say, "Socrates was a man, all men are mortal, therefore Socrates is mortal," there is no reason to ask myself whether this syllogism was socially determined. In contrast, it makes sense to ask why logic should choose an example concerning human mortality to expose this syllogism. This example tells me that during the period in which the phrase in question was pronounced, logical and moral philosophy were practiced by one and the same individual named "the philosopher."

These distinctions show that only observational errors and logical flaws, and axioms, assumptions, examples, and *ad hoc* hypotheses can legitimately form the object of a sociological analysis. The first category of factors—errors and logical flaws—pertains to the classical sociology of the sciences. The second—axioms, assumptions, examples, and *ad hoc* hypotheses—shows that an analysis of scientific content can be envisaged even within a rationalist framework. It should be specified notwithstanding that not just any scientific statement is subject to just any type of influence. This analytical approach based on the division of problems into classes is immune to criticism, because it focuses exclusively on the ascientific propositions that contribute to the construction of a scientific theory.

2) An opposition has been created between rationalists and relativists concerning the historicity versus the ahistoricity of the norms of rationality. And yet, in this case which unfolded in an era before the institutionalization of science, and where one observes three acceptations of the word "truth," science seems to have exclusively followed the norms of objectivity (the correspondence theory of truth) and rationality (the consistency theory of truth). This fact supports the ahistoricity of the norms of rationality. It shows that during the Middle Ages men of science were perfectly willing to reject supposedly authoritative arguments when the question being posed could be resolved through logic or experimentation. This hierarchy between the three norms—which was already recognized by Bacon—explains why one often finds the same procedures for deciding between what is true and what is false recurring in very different sociohistorical contexts.

One may find surprising the inordinate importance that certain sociologists and philosophers ascribe to the critique of the norms of rationality. The position of Stengers—that Reason should be substituted by History (Stengers 1993, 84)—stems from a previous text in which she argues that science is a "tradition" and that "all innovative work runs the risk of modifying this tradition, and the definition of rationality which it carries" (1988, 62). One may object that innovative research does not transform the *norms*, but the *statements* of science. No one can deny that scientific theories are subject to change. Even so, the transformation of scientific statements and theories is one thing; the transformation of the norms implicated in the construction of scientific statements and theories is quite another. If one fails to make this distinction, one is destined to confuse the substance with the form, and to equate rational statements (which in fact may change with every important scientific discovery) with the norms of rationality (which are not subject to such changes). The rules of syllogism and the *modus ponendo ponens* have not lost their validity in today's world, however saturated it may be with computers and cyberfantasies.

And it was precisely by refusing to recognize these distinctions that the practitioners of cognitive relativism succeeded in obliterating the differences between science, myth, and politics: *omnia in omnibus et econverso*. This position is contradicted not only by other documented controversies which, like the medieval debate over optics, arrived in the end at a rational conclusion. The relativist position clashes equally with the age-old *retorsion argument*: "If truth is an illusion, on what grounds should I believe the sophist who assures me that it does not exist?" (Aristotle, *Metaphysics*, 4 and 11; Thomas Aquinas, *In 4 Metaphysica*, lectiones 6, 7; *In 11 Metaphysica*, lectio 5). This method of reducing an adversary to silence (*et cessabit disputatio*) is so ancient that it is surprising to find contemporary philosophers and sociologists who are still unaware of its existence.

Limiting ourselves to the question of scientific controversies, the critique aimed at relativist sociology of science admits of three different logical extensions:

1) Scientific controversies *sometimes* are resolved rationally (extramissionism versus intromissionism) and *sometimes* achieve closure through extrascientific factors (Pasteur versus Pouchet, according to Farley and Geison, Mendelsohn, Latour, and Collins and Pinch—a thesis that is in fact very difficult to support). This leads to the hypothesis of heterogeneity in scientific practices and particularly in the settlement of controversies.
2) Some controversies may be socially or politically determined, but these are *anomalies* that reveal either a dysfunction in the scientific institution or a specific underlying context for the production of the science in question (e.g., the Mendelians versus the Lyssenkists). This is to hypothesize that the existence of social influences on scientific content represents the exception rather than the rule.
3) The conclusion that a controversy has been determined by extrascientific factors is an *illusion* arising out of the interpretive categories that one applies to these phenomena (Pasteur and Pouchet, in the analysis of Roll-Hansen and Gálvez). This leads to the hypothesis of the failure of relativism and constructivism, a failure that itself calls for an explanation.

The controversies studied so far—Heterogenists versus Homogenists, Organicists versus Vitalists, Intromissionists versus Extramissionists—allow for any one of these three interpretations. All the same, the arguments accumulated make hypothesis 3 the most likely, that is that the spectacular results obtained by the SSK studies depend heavily on the categories of thought that are applied—or projected—onto scientific activity. The bias of SSK studies consists primarily in their claiming results that do not follow from their premises. Their interpretation is distorted: *primo,* by a contextual method of analysis that is approximative, confounding fortuitous connections and necessary relationships, and *secundo,* by the lax use of words such as "determination," which could create the impression that material relationships exist where in fact there are only superficial resemblances; "translation," "enrollment," which might convey the idea that you are referring to an object or a mollusk (one may recall Callon's study (1981) of the dispute over the fishing of *Pecten maximus* in Brest); or "convention," under whose banner one may confuse the acceptance of a procedure with the acceptance of a generalized state of the world.

The sociology of scientific controversies can accommodate these critiques, because it is possible to envisage an analysis of the content without being forced to pose the axiom of a historic fluctuation in the norms of rationality. As this chapter has demonstrated, the link between these propositions arises out of a *historical contingency*, that is to say, out of the relativist impulse given to SSK by the Edinburgh school. If the eventuality of this tie is acknowledged, then the rationalist program of SSK may be reinforced by the addition of two principles:

—The investigation must detach itself from a simplistic conception of rationality, and distinguish at least between the *norms* of rationality and the rational *statements* that are constructed on the basis of these norms;

—Once we have defined the norms of rationality for the production of science that prevail in a given local context, the inquiry must determine, in each particular case, the *conformity* of the scientific statements with these norms.

Notes

1. It is necessary to make a distinction between the methodology and the results, because the conventionalized nature of a method does not at all connote that it will lead to a conventional result. *There is a gap between the nature of one and the other.* See, for example, the different ways of carrying out multiplication that were developed in different cultures (Chapter 6).

2. "But one alone knows the sciences, the Bishop of Lincoln"; "Sed solus unus scivit scientias ut Lincolniensis episcopus" (*Opus tertium*, X, ed. Brewer 1859, 33). "And Master Robert, called 'grosse teste', alone knew the sciences. For very illustrious men have been found, like Bishop Robert of Lincoln and Friar Adam de Marisco, and many others, who by the power of mathematics have learned to explain"; "Et solus dominus Robertus dictus grossum caput novit scientias. Inventi enim sunt viri famosissimi ut episcopus Robertus Lincolniensis et frater Adam de Marisco et multi alii qui per potestatem mathematicae sciverunt causas omnium explicare" (*Opus majus*, III, *hujus persuasionis*, IV, I, 3; ed. Bridges 1964, 108).

3. "Si adfectus videndi imaginibus adfluentibus efficeretur et ab omnibus corporibus perpetuo imagines effluerent, quae sensum nostrum adficerent, quaenam causa est, cur is, qui acum quaerit librumque perlustrat acum omnesque litteras non conspiciat?" (*Opticorum recensio Theonis*, X; ed. Heiberg 1895, 149–151).

4. "Catus tamen dicitur habere tantum de lumine ut totaliter sufficiat medium disponere et objectum et puto verum esse." (*Tractatus de perspectiva*, IV, 89; ed. Lindberg 1972, 39).

5. "Visum non fieri nisi per medium dyaphonum [. . .] Nulla densitas prohibet omnino transitum virtutum et specierum, quamvis non lateat. Hinc linces videre dicuntur per medium parietem" (*Perspectiva communis*, I, 51, ed. Lindberg 1970, 132).

6. "Causa vero acuminis visus in videndo a remotis est propter situm oculorum, nam oculus prominens non videbit bene a remotis. Et oculus profundus

videt remote, nam motus eius non dividitur neque consumitur, sed exit ab eis virtus visibilis et vadit recte ad res visas" (*De animalibus*, XIX, 781a; van Oppenraaij 1992, 220).

7. "Invenimus visum quando inspexerit luces valdes fortes fortiter dolebit ex eis et habebit nocumentum, aspiciens enim quando aspexerit corpus solis non poterit bene aspicere ipsum, quoniam visus eius dolebit propter eius lucem" (*De aspectibus*, I, 1, 4.1; ed. Smith 2001, 3).

8. "Relinquitur igitur quod lux multiplicatur in tempore et omnis species rei visibilis et visus similiter" (*Opus majus*, V, I, IX, 3; ed. Lindberg 1996, 138).

9. "All things that are seen require time for their perception. [This is so] because change becomes sensible only in a period of time, as we are taught by illusions of the senses in the swift movement of things"; "Omnia que videntur tempore comprehendi. Immutatio enim sensibilis non fit nisi in tempore, sicut docent illusiones sensuum in veloci quorundam transportatione." (*Perspectiva communis*, I, 53; ed. Lindberg 1970, 134).

10. "Ergo non est necesse ut radiis quasi nuntiis requiratur. Amplius quomodo aliqua virtus oculi usque ad sidera protendetur etiam si corpus totum in spiritus resolveretur?" (*Perspectiva communis*, I, 45; ed. Lindberg 1970, 128).

11. "Sententiat ergo Alfraganus ex comparatione semidiametri terrae ad semidiametrum orbis stellati quod distantia orbis stellati a centro terrae est 65.357.500 milliaria quod si duplicetur erit diameter totius orbis stellatis scilicet 130.715.000 milliaria." (*Opus majus*, IV, III, 6). As the author writes a few pages earlier: "Oportet igitur suponere quod milliare continet 4000 cubitorum" (one *cubitus* is about 50 cm).

12. "Unde philosophi naturales tangentes id quod est ex parte visus naturale et passivum, dicunt visum fieri intussuscipiendo. Mathematici vero et physici considerantes ea quae sunt supra naturam tangentes id quod est ex parte visus supra naturam et activum dicunt visum fieri extramittendo. Hanc partem visus quae fit per extramissionem exprimit Aristoteles aperte in libro de animalibus ultimo dicens 'oculus profundus videt remote nam motus eius non dividitur neque consumitur sed exit ab eo virtus visualis et vadit recte ad res visas'. Et iterum in eodem 'tres dicti sensus scilicet visus auditus olfactus exeunt ad instrumentis'. Perspectiva igitur veredica est in positione radiorum egredientium" (*De iride*, 73, ed. Baur 1912, 73).

13. "Tamen non est confusio istarum specierum, nec mixtio, nec fit unum ex eis, quoniam non sunt eiusdem speciei, nec eiusdem generis" (*Opus majus*, V, I, VII, 4, ed. Lindberg 1996, 106).

14. "Operationem visibilis in visum esse dolorosam. Hoc probatur quoniam operatio visibilis in visum est unius generis. Cum ergo operatio fortiorum lucium in visum sit lesiva sensibiliter et dolorosa sequitur omnes lucium operationes tales esse quamvis non perpendatur" (*Perspectiva communis*, I, 43; ed. Lindberg 1970, 126).

15. "Mathematicos ponentes visum fieri per radios ab oculos micantes superflue conari. Ergo superfluum est ponere sic radios" (*Perspectiva communis*, I, 44; ed. Lindberg 1970, 126).

16. "Sic ergo patet quoniam aliquo modo fit emissio radiorum sed non modo Platonico ut radii ab oculo emissi quasi in forma visibili immergantur et intincti revertantur oculo nuntiantes" (*Perspectiva communis*, I, 46; ed. Lindberg 1970, 128).

17. "Et ideo omnia praedicamenta dependent ex cognitione quantitatis de qua est mathematica et ideo uirtus tota logicae dependet ex mathematica." The following proposition opens Chapter 3: "In quo probatur per rationem quod omnis scientia requirit mathematicam [. . .] 8° In mathematica possumus devenire ad plenam veritatem sine errore et ad omnium certitudinem sine dubitatione quoniam in ea convenit haberi demonstrationem per causam propriam et necessariam. Et demonstratio facit cognosci veritatem" (*Opus majus*, IV, I, 2–3; ed. Bridges 1964, 105).

18. This theory offers a simple way to overcome the obstacle touching on the incommensurability between facts and statements (Lakatos 1980).

19. "Duo enim sunt modi cognoscendi scilicet per argumentum et experimentum. Argumentum concludit et facit nos concedere conclusionem sed non certificat neque removet dubitationem ut quiescat animus in intuitu veritatis nisi eam inveniat via experientiae" (*Opus majus*, VI, I; ed. Bridges, 1964: 167).

20. "Lux primum per oculos sola diffunditur emicatque in radiis oculorum ad visibilia contuenda" (*De genesi ad litteram*, XII, 16, 32).

21. "[Augustinus] nam vult in sexto Musice quod species visus vegetatur in aere usque ad rem" (*Opus majus*, V, I, VII, 2; ed. Lindberg 1996, 100).

22. "For the first time, the mirror cost 60 pounds of Paris, which are worth about 20 pound Sterling, and then I had a better one made for 10 pounds of Paris, and later, fully experienced in these matters, I realized that they could be better for only two shillings and 20 marks or even less"; "Primum enim speculum consistit 60 libris parisiensium que valent circiter 20 libras sterlingorum et postea feci fieri melius pro 10 libris parisiensium [. . .] et postea diligentius expertus in his percepi quod meliora possent fieri pro duobus marcis vel 20 soldis et adhuc pro minore" (London, British Library, Royal MS. 7 F VIII, fol. 4).

23. "Si igitur lucis multiplicatio est in instanti et non in tempore erit instans sine tempore quia tempus non est sine motu. Sed impossibile est instans esse sine tempore sicut nec punctum sine linea. Relinquitur igitur quod lux multiplicatur in tempore et omnis species rei visibilis et visus similiter. Sed tamen non in tempore sensibili et perceptibili a visu sed insensibili [. . .]" (*Opus majus*, V, I, IX, 3; ed. Lindberg 1996, 138).

24. "Omnia que uidentur tempore comprehendi. Immutatio enim sensibilis non fit nisi in tempore sicut docent illusiones sensuum in ueloci quorundam transportatione." (*Perspectiva communis*, I, 53; ed. Lindberg 1970, 134). One may note that Grosseteste on the contrary supported the thesis of the instantaneous propagation of light. At least this is what the translation would lead one to understand: "A point of light will produce *instantaneously a sphere of light [. . .] by multiplying itself *instantaneously in every direction [. . .]" The Latin text simply reads: "[. . .] a puncto lucis sphaera lucis quamuis magna subito generatur [. . .] seipsam multiplicando et in omnem partem subito se diffundendo [. . .]" (*De luce*, 51), a passage in which the word *subito* could equally well be translated "as soon as," thus diminishing its weight as an argument for the thesis.

25. "Sursum enim ad ipsum usque aethera et usque ad celum pervenit, in latitudine vero omnes mundi partes in brevi temporis momento illuminabat [. . .] Talis est enim ipsius natura tenuis scilicet et pellucida, adeo

ut lux per ipsum transiens nulla temporis mora indigeat" (*Hexaemeron*, homilia II, 7).

26. Galileo's experiment consisted in placing two men, each one holding a lighted lantern, a certain distance apart. At a given moment one would conceal his candle, and the other was instructed to do the same as soon as he saw the light of the other disappear. This experiment was conducted at varying distances, beginning with several *coudées* (a coudée is about 50 cm), then two or three miles (5 km), then eight or ten miles (15 km), observing the concealment of the candle with the aid of a telescope if necessary. If the timing of the covering of the two candles was identical regardless of the distance separating the two observers, this would signify that the propagation of light was instantaneous. If the timing varied as a function of the distance, this would constitute one proof that light was propagated over time. Galileo admitted: "Really I have tried it only at a small distance" *(Discourses,* I, 88). Therefore he decided to resolve the question intuitively, based on the analogy of the propagation of a flash of lightning: "Which seems to me a strong argument, that is some small time in doing, for had the illumination been made all at once, and not gradually, I can't think we could be able to distinguish its rise [. . .] and extremities of their dilatation" *(Discourses,* I, 89).

27. The rejection of the thesis of extramission was developed here on two fronts, which has certainly not been true for all theories; there have cases of massive adherence to false theories, and true theories that enjoyed little consensus.

28. The authors identify three modes of revision: *updating,* in which knowledge is adapted to fit objective changes in the underlying framework; *focusing,* in which a portion of the knowledge is transformed when the overall theory is unable to explain a particular case; and *revising,* in which knowledge is rectified each time that what one learns from the real world is in disaccord with the original theory.

29. The development proposed regarding mutually exclusive theories is logically irreproachable, but the cases to which it may be applied are without a doubt rather few in number. The author observes regarding the chosen example: "It must be added that the Earth, in fact, is not *truly* round (in this way, it is flattened at the poles). The theory [. . .] can therefore be replaced by a better theory and there is no limit to the precision with which one may describe its form (Boudon 1990, 149).

5

Al-Samarqandī's Native Theory of Controversies: An Essay on the Negotiation of Truth[1]

From an epistemological point of view, scientific activity consists in resolving enigmas by means of theories that will be accepted if they are accredited by empirical and rational proofs. A sociology of the sciences that takes into account the lessons of epistemology therefore cannot ignore the probationary component of scientific activity. And yet the moment one envisages the accompaniment of the logicist description by a sociological description, an asymmetry emerges. In fact, according to many contemporary authors, scientific knowledge does not present any particularities that distinguish it from ordinary belief. The ordinarism—or antidifferentiationism—of knowledge is one of the central theses of relativism and constructivism.

As we have seen in the previous chapters, there are various facets to this current of SSK. I will therefore allude frequently to the ensemble of relativist theses in my examination of just one of them: the "negotiation of truth."[2]

In defending this thesis, constructivist sociologists find themselves in an awkward position vis-à-vis epistemology, according to which one should decide between opposing hypotheses on the basis of experimental tests and a rational critique, which form the very underpinnings of the normative structure of science. From an epistemological point of view, in fact, neither the hypotheses nor the rational proofs that accredit them can be negotiated.

The aim of this chapter is to reduce the dissonance between the sociological and the epistemological conceptions of truth by examining the case of scientific controversies, which have often received particular attention from those seeking to prove that science is conducted through

163

negotiation and persuasion (see Latour and Callon 1991, 26). Indeed, during the course of a controversy, truth is temporarily suspended and replaced by an exchange of arguments of uncertain statute, making it an easy matter for the sociologist to show the hesitations and uncertainties that always accompany the process of constructing scientific knowledge.

The constructivist sociology of the sciences has often referred to the difference between rationality and irrationality as the "Great Divide," presenting a model of knowledge in the form of a map of the world in which Europe, the United States and Canada are a block of the same color while the rest of the world is left white and blank. The fact that scientific activity is primarily concentrated in Western countries means that the number of countries producing "scientific knowledge" will be far outnumbered by those producing "local beliefs." Thus, it can be posited that universal knowledge is contextual and contextual knowledge is universal. Such a map raises serious questions, however. Its bases have never been clearly explained because the map is supposed to be drawn by others scholars, and every attempt to reconstruct it using objective criteria (from the distribution of Nobel prizes to the percentage of GDP allocated for scientific research) fails. Finally, as every student of history knows, even in those zones designated by SSK as "producers of local beliefs," a tradition of scientific inquiry has always existed.

We may nevertheless attempt to follow the SSK thesis as far as it will take us by exploring how the "producers of local beliefs" perceived the concept of scientific controversy in a particular case—that of Arab scholars during the golden age of Islamic thought. Our analysis will center on the theory set out in the scholarly disputation, *Risāla fī adāb al-baḥth* by the mathematician and astronomer Shams al-Dīn al-Samarqandī (ca. 1250–ca. 1302). His writings will be taken as the departure point for an examination of the thesis of the "social negotiation of truth."

Samarqandī was certainly not the only savant to have contributed to an "internal epistemology" of scientific controversies. In the seventeenth century the German mathematician Gottfried Wilhelm Leibniz (1646–1716) was involved in polemics with, among others, René Descartes, Isaac Newton, Jean Bernoulli, Denis Papin, and Georg Ernst Stahl, and he left copious notes on "the art of controversies." Leibniz's thoughts on this subject are familiar to us thanks to the work of Marcelo Dascal et al. (2006), whereas Samarqandī is little known in the West even though his priority is indisputable, as he was the first author to have proposed a theory of scientific controversies.

This chapter will be divided into three parts. In the first I will explore the political model of the negotiation of truth that forms the basis for contemporary SSK analyses of scientific controversies. In the second part I will present Shams al-Dīn al-Samarqandī's theory of scholarly disputation and the juridical model on which it was based. In the third I will compare the political model espoused by contemporary sociologists with the theory of the Persian thinker in order to gauge the distance that separates these two conceptions.

In the light of this comparison, we will propose an alternative approach to writing about scientific controversies. The existence of a robust native theory that owes nothing to the idea of the social negotiation of truth allows us to envisage a *continuum* between the philosophy and the sociology of science rather than assuming an opposition between two mutually exclusive descriptions of scientific activity.

1. Science, Politics, and Negotiation

The contemporary sociology of the sciences provides a framework for the study of scientific controversies articulated in the form of various analytical programs, as we have seen in Chapter 1. According to the Strong Programme (Bloor 1976), scientific statements are causally determined by the state of society at the time these statements were constructed. In the Empirical Programme of Relativism (Collins 1981), the stabilization of scientific statements can be explained by sociopolitical structures that exert their influence *not* on the scientific content directly, but on the social procedures that limit the interpretative flexibility of the natural world. The Actant-Network Theory (Callon and Latour 1991) asserts that the stabilization of scientific statements represents the outcome of a battle between the groups mobilized by concurrent hypotheses, which will always work to the advantage of the one with the most extensive and powerful network.[3]

Taking as its departure point one or another of these programs, sociology has explored the broad spectrum of interests underlying scientific knowledge. The thesis of the negotiation of truth originated out of its examination of one of these factors—the political dimension of scientific activity. Based on the assumption that the scientist can never provide *definitive* proof of a hypothesis, it asserts that he must attempt to gain acceptance for what he believes by using his power or by negotiating compromises with his equals. In a scientific controversy, therefore, the researcher is fighting with the same weapons as the politician on the battlefield of public policy. Therefore the Actant-Network

Theory has been considerably influenced by the model of the negotiation of truth.

The political model is central to the analysis of scientific controversies, first of all because according to SSK the closure of a controversy can be ascribed to an asymmetry between the victor and the vanquished, in which the winners prevail not necessarily because they are right, but because they benefited from greater support: "A Great Divide between scientific minds does not exist, only networks, some of which are longer or shorter than others" (Latour 1989, 428). The political model is applicable therefore, because the negotiations in which the scientists are presumed to be engaged directly corresponds to the paradigm of the man of politics (Latour and Woolgar 1988, 253, 261).

Other authors equally inspired by the constructivist school have sought to explain the functioning of scientific activity by drawing a link with the world of politics; for example, Shapin and Schaffer (1989), and Karin Knorr-Cetina who concludes: "No interesting epistemological difference could be identified between the pursuit of truth and the pursuit of power" (1995, 151).

It was Berthelot who in 2008 proposed a fresh approach to the sociology of the sciences by reintroducing the idea of *truth* into the process of the production of scientific knowledge. His thesis can be summarized in five axioms, of which the third is the differentiation of the spheres of activity. Although his position is far removed from that of the constructivists described above, Berthelot allows a place for the political model when he writes: ". . . the scientific sphere differentiates itself by the introduction of acceptation criteria issuing from the political sphere" (Berthelot 2008, 209). This idea was drawn directly from the work of the historian of Greek science, Geoffrey E. Lloyd although, if we return to the original source, we find that Lloyd's reflections on this point are parenthetical:

> Chinese intellectuals do not regularly undertake the total demolition of their rivals' views by way of undermining their epistemological and methodological assumptions [as instead the Greeks did . . .]. Where Greek polemic so often adopts models influenced by the *law courts* and *political assemblies,* Chinese advisers often envisage a situation of persuading the person whose opinion really counted, namely the ruler (Lloyd 2003, 55, italics mine).

However, Lloyd raises another pertinent point that has hitherto been overlooked. The political model is not the only one that can be applied to the analysis of scholarly disputes. He noted that the construction of a dispute can be compared to the process of arguing a case before a court of law. That the successive stages of a court case, which is based on the search for evidence and the truth, may serve as a model for the scientific debate should come as no surprise. What is surprising is that this filiation has never been addressed before by historians and sociologists of the sciences. We are therefore justified in asking whether the broad acceptance of the thesis of the negotiation of truth may not obscure alternative models that could be equally valid in the analysis of scientific controversies.

2. Samarqandī's Theory of the Scholarly Dispute

The Persian savant Shams al-Dīn al-Samarqandī (ca. 1250–ca. 1302) is known for his work in logic, mathematics, and astronomy. In particular, we owe to this master of *kalām* the *Tadhkira fī'ilm al-hay'a (Synopsis of Mathematical Astronomy)* and a star catalogue, *'Amāl al-taqwīm li-al-kawakīb al-thābita (Construction of the Calendar of Fixed Stars)*, which dates to the year 1276–77. Another important contribution was his *Kitāb ashkāl al-ta'sīs (Book of Fundamental Theorems)*, a commentary on thirty-five propositions from Euclid's *Elements* (Dilgan 1960, 1975; Rozenfeld and Yushkevich 1961; Suwaysī 1984; De Young 2001; Fazlıoğlu 2007).

By defining the procedural rules to be followed in the conduct of a scholarly controversy, Al-Samarqandī invented an entirely new discipline. It appears that he wrote four treatises on the subject, of which three have survived: *Risāla fī adāb al-baḥth (Epistle on the Rules and Etiquette of Debate)*, *Qusṭās al-afkār (The Weighing of Ideas)* and *al-Muʻtaqadāt (The Convictions)*. Only the last text, *al-Anwār (The Lights)*, has not come down to us.[4]

Few facts are known about the life of al-Samarqandī. Around 1268 he was studying under the eminent jurist Burhān al-Dīn Muhammad ibn Muhammad al-Nasafī (d. 1288). Other than that, a few inferences can be drawn from the order of appearance of his works. *Risāla* predates *Qusṭās*, and *Qusṭās* was written shortly before 1291, the year in which a first copy of it appeared. It can be deduced that *Qusṭās*, *al-Muʻtaqadāt*, and *al-Anwār* were written before *Sharḥ al-muqaddima al-burhāniyya*, because the latter contains citations from all three of these works. The four treatises by al-Samarqandī on the art of scholarly disputation were therefore composed in the period between 1268 and 1302. The relatively early date of his *Calendar of Fixed Stars* (1276–77) strongly suggests that his reflections on the art of the dispute came after his scientific work, but for the moment there is nothing to exclude the possibility that the former were written before or in parallel with his scientific treatises.

1. Samarqandī's Interest in the Nature and Resolution of Controversies

It might appear paradoxical that Samarqandī should have written four essays on a juridical conception of the scholarly dispute when, in his own hierarchy of the sciences, law was placed on a lower rung than the exact sciences such as logic, mathematics, and astronomy. Why then did Samarqandī take such a close interest in this discipline? Did he become embroiled in polemics that caused him to turn to the juridical dialectic for a solution regarding how to conclude a dispute in a fair manner?

Questions of mathematics and astronomy regularly formed the subject of lively polemics in the Arab world beginning in the ninth century, perhaps due to the popularity of critical treatises composed in accordance with the conventional format of *Sharḥ (Elucidation), Tanqīḥ (Rectification),* and *Shukūk (Queries).* During this time, a series of controversies arose as Arab intellectuals began to challenge the Ptolemaic model (Tahiri 2008). Other discussions sprang up over the construction of the regular heptagon, the trisection of an angle, and the duplication of the cube. Some of these debates became quite celebrated, such as the one that opposed Abū al-Jūd, Ibn Sahl and al-Sijzī regarding the construction of the regular heptagon by means of conic sections (Hogendijk 1984; Rashed 1999, 647–898), and the eighteen *Questions and Answers* in physics, astronomy, and cosmology which was the focus of a long exchange between al-Bīrūnī, Ibn Sīnā and his disciple Abū Saʿīd al-Maʿsūmī (Berjak and Iqbāl 2003). The scholar al-Bīrūnī himself wrote about the discussions leading up to the establishment of the sine theorem (Debarnot 1985).

Samarqandī could not have remained unaware of these polemics and he seems to have become involved in a few scientific controversies himself. In examining the thirty-five propositions of Euclid's *Elements,* he did not hesitate to criticize Ibn al-Haytham, ʿUmar Khayyām, and two of his contemporaries—Athīr al-Dīn al-Abharī (d. 1265) and Naṣīr al-Dīn al-Ṭūsī (1201–74)—who defended opposing theses regarding the bases of the propositions under discussion. Samarqandī's *Kitāb ashkal al-ta'sīs* therefore assumed the contours of a mathematical dispute. It is plausible that the frequency of scientific controversies during the golden age of Arab science and Samarqandī's participation in one of them inspired him to undertake his philosophical-juridical studies into this subject.

Samarqandī maintained that his theory of scholarly disputation could be applied to any discussion between two parties who found themselves in *disagreement* over a question regarding which one could arrive at the *truth*. All of the sciences therefore could make use of his theory of controversies, particularly the fields of philosophy, logic, astronomy, and mathematics (*Qusṭās*, fol. 59a; Miller 1984, 206). His work had a great influence on successive authors. In fact, the *adāb al-baḥth wa-al-munāzara* became a distinct genre in Arab literature, and was inherited by the Ottomans who produced many illustrious practitioners including Aḍud al-Dīn al-Ījī (d. 1355), al Sayyid al-Sharīf al-Jurjānī (1339–1413), Taşköprüzāde (d. 1561), al-Marʿashī (d. 1737), Saçaklızāde (d. 1737), and Gelenbevī (d. 1791) (see Miller 1984; Karabela 2010).[5]

2. Samarqandī's Juridical Model

Samarqandī's crucial contribution was to redefine the debate center-ing on contradictory theses, which was known among Arab logicians as "a discussion between two sides in order to reveal the truth. If it is not done to reveal, it is dialectic (*mujādala*)" (*Sharḥ al-muqaddima al-burhāniyya*, fols. 40b–41b; Karabela 2010, 125). In his redefinition, Samarqandī separated the rules of scientific disputation (*adāb al-baḥth*) from the rules of dialectic (*adāb al-jadal*), the requirement for the truth operating as the criterion of demarcation. Here Samarqandī must be credited with "logicizing" the traditional Arab dialectic (Rescher 1964, 209).

However, even in this rupture with the past, the sources that Shams al-Dīn al-Samarqandī drew upon to conceive his theory can be imme-diately detected. The first source of *adāb al-baḥth* was Aristotle, who defined the principles of the oratory joust in Book VIII of his *Topics* (Moraux 1968). Likewise, the four fundamental questions of the debate, which were assimilated by all Arab philosophers, had an Aristotelian origin: "We seek four things: (1) the fact, (2) the reason why, (3) if something is, and (4) what something is" (*Posterior Analytics*, 89b24). This text was well known to medieval Islamic scholars.

The Arab dialectic detached itself from Aristotle and became more autonomous when scholars began to apply it to different spheres of activity such as theology, law, and philosophy. Theoreticians then turned their attention to aspects that Aristotle did not consider, such as how a debate commences, how it ends, and the manner in which it is conducted (cf. Aristotle, *Topics*, Book VIII). Thus Fārābī (ca. 872–950) described how O (the opponens) receives the thesis and premises of R (the respondens), how O may refute them, and how R may challenge the form of O's syllogism. Depending on the specific case, either O or R may produce the *elenchus* (al-Fārābī, *Kitāb al-jadal*; Miller 1984, 78–80).

All the same, it can be observed that Samarqandī's philosophical authorities were eclipsed by his juridical sources due to the highly sophisticated development of the dialectic of truth by Arab scholars. They freed the dialectic from its Greek roots by applying it to the law (*uṣūl al-fiqh*) and to jurisprudence (*furū'*). The principal authors of this current were al-Bājī (1013–81), al-Shīrāzī (d. 1083), al-Juwaynī (1028–85), and Ibn 'Aqīl (1140–19). This emancipation was most visible in the replacement of the syntagm *adāb al-jadal* (the rules of dialectic) by *adāb al-khilāf* (the rules of controversy). This led to new questions that complemented and completed Aristotle's four original questions, such as: How can the respondent circumvent the refutation (al-Juwaynī)?

In the same way, the theory of *jadal* underwent a series of novel developments when it was applied to the law. Let us analyze just one example. The Mālikite jurist Abū al-Wālid al-Bājī (d. 1081) identified three broad types of objections that might be addressed by the disputant to a thesis: the request (*muṭālaba*), the objection proper (*i'tirād*), and the counterobjection (*mu'āraḍa*). Later authors would add other objections, so that by the time of Samarqandī the contradictor had no less than ten strategies with which he might attack an adversary. Here is the list:

1. *Request (muṭālaba)*. The request is made by O, who asks for an explanation or verification of R's proof.
2. *Objection (i'tirād)*. The objection of O must address the crux of the proof.
3. *Counter-objection (mu'āraḍa)*. The counter-objection of R consists in opposing the objection of O with an objection of even greater force.
4. *Disallowance (mumāna'a, man')*. The disagreement here focuses on a characteristic inherent to the case under discussion that prevents the application of reasoning.
5. *False construction (fasād al-wad')*. A false construction is present when R's argumentation contains a logical flaw.
6. *Ineffective ratio legis ('adam al-ta'thīr)*. O can object that R is linking a cause to a principle on the grounds of a relationship that does not exist.
7. *Incompatible conclusion (qalb)*. O can accept the demonstration proposed by R, but nevertheless draw from it a different conclusion.

8. *Inconsistency* (*naqḍ* or *munāqaḍa*). An inconsistency is present if the argument advanced is not valid or if the reasoning is self-contradictory.
9. *Limited validity* (*al-qawl bi-mūjib al-'illa*). O can object that the principle invoked by R, while valid in itself, cannot be applied to the case under examination.
10. *Distinction* (*farq*). O can introduce a distinction if either the main case or the case under discussion appear to him to arise from different principles (for a discussion of this, see Miller 1984, 109–134).

Neither Aristotle nor the early Arab dialecticians had at their disposal such a complete range of potential strategies. Consequently, the genre for which Samarqandī was proposing his radical synthesis and reelaboration was no longer *adāb al-jadal* (art of dialectics), but a new *adāb al-khilāf* (art of disputation)—a set of rules for the conduct of a debate that could only have developed out of the author's contact with Arab law (*uṣūl al-fiqh*) and jurisprudence (*furū'*). The juridical origin of his rules are clearly acknowledged by Samarqandī who nevertheless, to explain the autonomous nature of his rules of debate, writes: "[. . .] the juristic dialectics (*khilāf*) of [his] times does not need it [the art of dialectics] any more" (*Qusṭās*, fol. 59a; Miller 1984, 200).

Why was Samarqandī inspired by the juridical model in his work on controversies rather than the more well established modalities of negotiation (*mufāwaḍa, mubāḥatha, musāwama*), which were as fully developed in the Arab world as they were in the Latin world? This can be explained by certain distinctive features of Arab dialectics that we will explore by comparing Samarqandī's juridical model with the political model for the negotiation of truth.

3. Two Antinomic Models

Samarqandī's *Risāla fī adāb al-baḥth* (*Epistle on the Rules and Etiquette of Debate*) is composed of three extended sections in which he presents: (1) Definitions of the relevant technical terms, such as disputation (*munāẓara*), proof (*dalīl*), objection (*man'*), etc. (2) An exposition of the order to be followed in the discussion (*tartīb al-baḥth*), in which he treats such questions as: What is a valid objection? How can one determine when a controversy has come to an end? (3) The application of these rules to different philosophical problems, as chosen by al-Samarqandī. Each of these sections contains elements comparable to those in contemporary models for the analysis of scientific controversies.

1. The Order of the Debate

Constructivism does not fix or follow particular rules regarding the order in which arguments should be exchanged during a scientific controversy. According to the tenets of SSK, the fact that there are no ethical rules to frame the debate explains why the negotiations undertaken by scientific researchers are opportunistic in character (Latour and Woolgar 1988, 261). In contrast, according to medieval Arab philosophers every controversy is a discussion based on norms, in the philosophical dialectic *(jadal)* as in the juristic dialectic *(khilāf)*. A scientific controversy is a debate (*al-munāẓara*) in which two disputants, each with a well-defined status, come face to face: the opponens O (*sā'il*) and the respondens R (*mujīb*). The contradictory debate is also normative by virtue of the fact that it is required to address in turn four questions (al-Maqdisī, *al-Bad'*, 50, French translation, 46):

1. The whatness of R's opinion (*mā'īyat al-madhhab*);
2. The proof of R's opinion (*dalīl*);
3. The cause (*al-'illa*) or necessary reason (*al-sabab al-mūjib*) for R's opinion;
4. The verification of the cause (*taṣḥiḥal-'illa*) of R's opinion.[6]

Samarqandī goes even further by envisaging the argumentative structure that can be mobilized, and the order in which the adversaries should be allowed to speak during the debate. The *Risāla fī adāb al-baḥth* and *Quṣṭās* provide similar expositions in this regard. Here is a passage from the *Quṣṭās*:

1. When R begins the disputation, it is incumbent upon him, before he establishes the proof for his claim (*mā idda'āhū*), that he (a) explain the objects of his investigation (*taḥrīr al-mabāḥith*) and (b) establish (*taqrīr*) the opinions and beliefs, so that the point of dispute (*ṣūrat al-nizā'*) becomes perfectly clear.
2. O may here demand a verification of the attribution (*taṣḥīḥ al-naql*) of the opinions and beliefs; for often defects occur in the debate since R might pretend to be arguing with someone other than his actual opponent and use premises granted by this other person as if there were granted by his actual opponent. This, however, leads to randomness in debate (*khabṭ*).
3. But when R begins to establish a proof for his claim, then O may either (i) object or (ii) not object. If he does not object then it is clear. If he does, he may do so either (a) before R is finished bringing his proof or (b) afterwards. If the former (a), then O may merely object (*yaqtaṣiru 'alā mujarrad al-man'*) or not. If not, he may do so with backing (*mustanad*) or not. . . .

4. But if (b) O objects after R is finished bringing his proof, he may either grant R's proof or not. . . . But if he accepts the proof, then he must reject that it proves R's point (*al-madlūl*) basing himself on some other piece of evidence or not. . . . All this advice is for O.
5. As for R, he must ward off any objection by bringing evidence (*dalīl*) or alerting O to something which is known a priori (*tanbīh*) . . . .
6. The debate continues until R is silenced (*ifhām*) or O is forced to accept his argument (*ilzām*) (*Qustās*, fol. 59b; Miller 1984, 210–211).

Because it is somewhat difficult to follow this reasoning, which takes the form of a logical tree, I present in Figure 4 a partial dichotomic tree corresponding to the objections that O might address to R.

To summarize, the rules enunciated by Samarqandī seek to preserve the unity of the debate, either by ensuring its completeness (with the request for complementary clarification or proofs) or by limiting its dispersion (by rejecting any digressions).

2. The Weapons of the Disputant

Constructivism does not detect any pattern of regularity in the arguments that are exchanged during the course of negotiations in a controversy. The success of one side, and as a consequence the force of the theories that it is defending, depend entirely on its persuasive capacities (Latour and Woolgar 1988, 65). The Arab dialectic does not conform to the constructivist description. According to SSK, the outcome of the debate will depend exclusively on the opportunist capacity of the researchers to convince their interlocutors. By comparison, Arab theoreticians seem to have been motivated by a keen interest in codifying the arguments that might be used in a controversy. Earlier treatises on *jadal* and *adāb al-khilāf* had already enumerated a number of strategies that could be utilized by the opposant. In *Risāla*, Samarqandī draws up a synthesis of these contributions and proposes an empirical classification of the resources of the disputant (Miller 1984, 234):

1. Contradiction (*munāqada*): O disallows a premise of R because it is false, and in this way refutes the conclusion that is drawn from it;
2. Counterobjection (*mu'ārada*): O develops a parallel demonstration, which is contrary to that of R;
3. Inconsistency (*naqd*): O objects that the demonstration presented by R for the case under discussion lacks legitimacy, that is legal status (*hukm*);
4. Backing (*mustanad*): O refuses the reasoning of R and produces a document external to the discussion, on which he bases his objection.

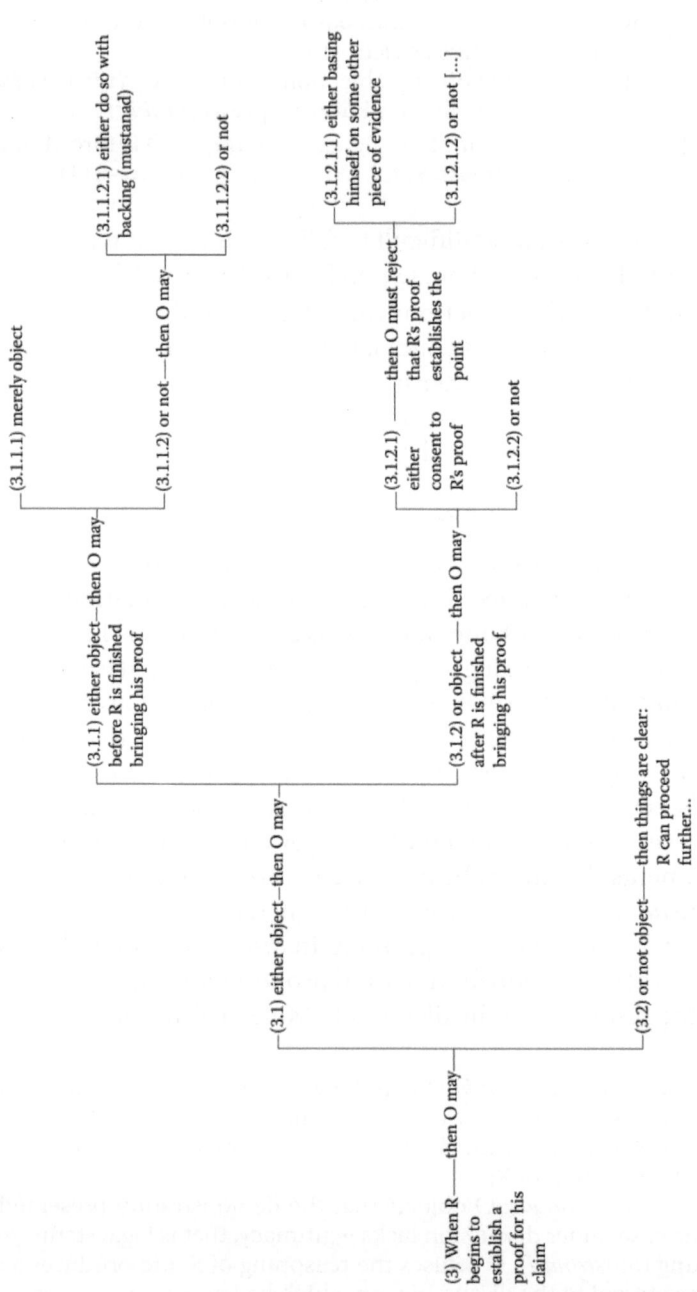

Figure 4. A dichotomic tree of O's objections.

By means of this orderly classification Samarqandī reduces the number of strategies at the disposition of the disputants to the four logical types outlined above. These four types lead us directly to the two classical conceptions of truth—the correspondence (1, 4) and the consistency theories of truth (2, 3). One of the most important aspects of Samarqandī's theory is that it recognizes and maintains the rational dimension of scientific activity, which has in the twentieth century been evaded by constructivism.

According to *adāb al-baḥth*, the practice of scientific debate resembles a stage performance or the *collective and contradictory work* of parties whose intention is to find out the truth on the basis of tangible proofs, more closely than the discussions in the corridors of power where compromises are negotiated. A comparison of the two models shows moreover that negotiation does not have as its scope the inclusion of the totality of social interactions. There is little room for it in the domain of the sciences.

3. The Settlement of Controversies

In the contemporary sociology of the sciences, the term "settlement" covers three quite different realities: resolution, if the controversy attains a rational solution through an internal process of development; closure if external factors are responsible for its settlement; and abandonment in the other cases (Mendelsohn 1987; McMullin 1987). For their part Engelhardt and Caplan (1987) distinguish up to five modes for the settlement of controversies: (1) through loss of interest, (2) through force, (3) through consensus, (4) through sound argument, and (5) through negotiation. Of these five modes, only the fourth envisages the possibility of the *rational* resolution of a controversy. In the political model of constructivism, this analysis is greatly simplified by reducing the modes that are actually operative to types 2 and 5, the ones judged to be the most preponderant. Thus, in a controversy the party with the most extensive and powerful network will prevail. One consequence of this simplification, however, is that it eliminates the possibility of a rational resolution of the debate.

Here again the Arab theory of debate distinguishes itself by its originality. The authors of *adāb al-jadal* (rules of dialectic) and *adāb al-khilāf* (rules of controversy) introduced an innovation aspect, which they denominated the "signs of defeat" (*dalā'il al-inqiṭāʿ*). These signs indicate at what moment and which party to the debate—either O or R—has emerged as the victor or the vanquished. Since the codification of the signs of defeat may differ slightly from one author to another, in the analysis of these lists it is important to include the names of the authors who may have referred to them. One will find side by side the authors of *jadal*, such as al-Ash'arī (874–936; denoted below as A) and al-Maqdisī (1160–1223; denoted M), and the authors of *khilāf*, such as al-Baghdādī (1002–1071; denoted B) and al-Juwaynī (1028–1085; denoted J). The defeat of one adversary by the other can be identified by the following signs, presented in descending order from the more consensual to the less consensual (Miller 1984, 39–40, 139–141):

1. Inconsistency (*naqd, munāqaḍa*) (A M B J). The defeat is manifest if O succeeds in showing that the reasoning of R is inconsistent, that is if the conclusion is not proportional to the premises or if it is self-contradictory.
2. *Reductio ad absurdum* (*qiyāsu al-khilfa*) (A M B J). O can demonstrate that the thesis of R is false, because logically it leads to an absurd consequence.
3. Silence (*ṣamt*) (A M B). It is surprising to find silence as a sign of defeat in its own right, since every argumentative strategy is aimed at reducing the adversary to silence, cf. the Latin formula *cedat tempus!*
4. Distinction (*farq*) (A B J). R is held in check if O can show that R has failed to make a distinction that is instead necessary to his chain of reasoning.
5. Incapacity (*al-sukūt ill-'ajz*) (A M). R is vanquished if he finds himself incapable of responding to a question posed by O. This is a variation of the reduction to silence described in (3).
6. Digression (*intiqāl*) (A M). Digression is a sign of defeat, because it marks a break in the continuity of R's reasoning (logical fault).
7. Commensurability (*al-tanāsub*) (A M). R must admit failure if the case examined does not conform to the case-type on which he has chosen to model his argument.
8. Deviation (*al-īnhirāf*) (A). R loses if he responds to a different question from the one that was posed. O merely has to underline the logical rupture.
9. Appeal to the crowd (*nidā' ilā al-jamāhīr*) (M). R has lost if he makes a direct appeal to his listeners; this signifies that he has no further arguments to offer.
10. Stubbornness (*mukābara*) (M). By the same token, R is defeated if he repeats his thesis and refuses to admit the (valid) objections that have been raised.

This classification system is of immense interest for the study of scientific controversies. For the disputants, signs of defeat are the *social and rational norms* that guide the manner in which one should behave in a contradictory debate (for details, see Vajda 1963, 7). If these norms are transgressed, the disputant is disqualified and the advantage is granted to his adversary. For all this, the social norms of scholarly disputation must not be idealized. On the one hand a historian such as al-Tawḥīdī (923–1023) describes them as a combination of sociability and courtesy aimed at guaranteeing equity (*al-tanāṣuf*) between the parties and dissipating any fear (*al-taqiyya*) of the one who expresses himself (al-Tawḥīdī 1970, 21). At the same time there are documents which testify to episodes of duels in "bad manners" (Stroumsa 1999, 68).

4. Conclusions

From this study it appears that the *adāb al-baḥth* presents a challenge to the contemporary school of constructivism. Four aspects deserve to be underlined. (1) The theory of Samarqandī proposes a genuine sociology of rationality. (2) The formulators of this theory never contemplated the possibility of the negotiation of truth. (3) The juridical model on which it was based recognized the distinction between the true and the false even as a discussion was still in course. (4) Samarqandī's theory has an original contribution to make to the analysis of the settlement of controversies. Let us elaborate on each of these points.

1. Samarqandī's Internal Epistemology

Does subscribing to the native theory of Samarqandī require that one abandon the sociology of the sciences in favor of what Piaget refers to as an "internal epistemology"? Not at all. A commentator on Samarqandī, 'Aḍud al-Dīn al-Ījī, writes: "Sword and spearhead both achieve what demonstration (*burhān*) cannot achieve" *(Mawāqif fī 'ilm al-kalām;* Karabela 2010, 42); in other words, "The contradictory debate produces a result that private reflection cannot attain." Al-Ījī therefore teaches us that activities involving the search for the truth are group practices. The exercise of rationality is a social activity when it is conducted in relation to another party, as in the case of a controversy between two disputants. As a consequence, the Arab theory of controversies does include aspects that are clearly sociological in nature: it admits the superiority of collective debate over individual reasoning in the search for the truth, and it assumes that the establishment of knowledge rests on collaborative activities. In this, *adāb al-baḥth* constitutes the vehicle for an authentic sociological theory of rationality, which is not so very remarkable in itself, but which anticipates the position of contemporary authors according to whom objective knowledge is the result of a collective critical enterprise (Bachelard 1972, 241; Popper 1979, 148–149; Berthelot 2008, 203).

Table 7. Protagonists in the debate on atomic weights (1815–1919).

Pro	Contra
Prout 1815–32	Berzelius 1819
Thomson 1818–25	Ure 1821
Phillips 1824	Turner 1829–33
Turner 1825	Stas 1849–60
Henry 1829	Meyer 1885
Dumas 1828–43	Morley 1895
Penny 1839	Aston 1919
Marignac 1842–46	
Erdmann and Marchand 1843	
Gerhardt 1845	
Maumené 1846	
Clarke 1892	
Rutherford 1911	
Moseley 1913	

2. Samarqandī's Indifference to the Negotiation of Truth

The theory of Samarqandī is indifferent to the thesis of the negotiation of truth and therefore to the objections that may be raised against the latter. The thesis that scientific statements are modeled by negotiation is only convincing if one studies controversies over the short term. It is therefore not surprising to find the SSK thesis being embraced by the ethnomethodology of science laboratories, which always operate in the short term. When one examines scientific activity over the long term the diagnosis is reversed.

Let us return for a moment to the contemporary world of science. The controversy over atomic weights that divided the world of chemistry between 1815 and 1919 furnishes us with an illustration (Lakatos 1980; Hamerla 2003; Nendick 2007; Brock 2010). In 1815 the English chemist William Prout proposed the hypothesis that the atomic weights of all the elements were integers, specifically multiples of the atomic weight of hydrogen—the base unit. Jean-Servais Stas sparked the debate after determining that the values in relation to hydrogen for many key elements such as carbon, oxygen and chlorine were not integers. The Belgian chemist, who succeeded in measuring the weights of these elements to a high degree of precision (C = 11.97, O = 15.96, Cl = 35.46), became one of the most determined adversaries of Prout's hypothesis. As chemists around the world became aware of the growing disagreement, they collectively seized upon the question. Accusations of "scientific fraud" and the peddling of "illusions" flew back and forth, but the debate continued for more than a century without either side knowing who was right. The majority of chemists continued to support Prout's thesis; indeed, his allies seem to have been both numerous and influential, whereas the opposing thesis garnered considerably fewer adherents, as the list of the principle actors in the controversy shows (Table 7).

Since the experimental values, when measured with precision, did not fit the thesis of simple multiples of a unit weight, attempts were made to reduce the discrepancy by allowing for fractions of the atomic weight of hydrogen, first one-half (according to the Swiss chemist, Jean Charles Galissard de Marignac) and then one-quarter (by the French scientist, Jean-Baptiste Dumas). Such *ad hoc* hypotheses were clearly conceived in order to salvage Prout's hypothesis. But science would have to await the discovery of the stable isotope by Francis Aston and Frederick Soddy before any weaknesses were admitted in Prout's argument. The enigma would not be cleared up until 1919, when Aston proposed the notion that isotopes of the same element might be present in nature in different proportions, as an explanation of the fractional values found earlier by Jöns Jacob Berzelius, Edward Turner and Jean-Servais Stas. Let us take the example of one common element, chlorine. Once it was established that chlorine has two main isotopes—^{35}Cl and ^{37}Cl—present in proportions of 75.8% and 24.2%, it could be easily deduced that the molar mass of chlorine is about $(35 \times 75.8 + 37 \times 24.2) / 100 = 35.48$.

For more than a century the hypothesis of Prout benefited from a "longer and more powerful network," but this circumstance fails to explain how the controversy was finally settled. The theory of Prout was simply discounted and quietly forgotten when chemists discovered the existence of stable isotopes.

Constructivism presents two blind spots. First, by assuming negotiation to be the driving force of the discussion, the constructivist thesis obscures one of the most important aspects of the researcher's work, which is to adjust scientific statements to fit the structures of the objective world. Second, by focusing on the themes of power and persuasion, constructivism ignores the fact that researchers demand objective proof before accepting a theory. In contrast, proofs and correct reasoning were cardinal principles that the Arab theory of *adāb al-bahth* did not renounce.

3. Samarqandī's Juridical Model

If Samarqandī could conceive an art of conducting scientific controversies based on the example of juridical procedure, it must be concluded that the contemporary model of the negotiation of truth cannot serve as a universal guide for the study of scientific controversies. There is a fundamental difference between two approaches here. The political model seeks to demonstrate that simple "scientific beliefs" are installed by a process in which shared resources, such as the capacities of persuasion and negotiation, are called upon in a collaborative social activity. Samarqandī's model seeks to delineate the course of a controversy by following the *intrinsic internal order* of the debate. Even if this approach is not immune to criticism,[7] it presents advantages over the SSK thesis that make it more useful in the analysis of scientific controversies:

1. Samarqandī's model conserves the distinction between the true and the false. But far from being detached metaphysical categories, the true and the false are directly linked to the practical approach in the debate; there are alternative paths that O and R can take, but some are false paths that lead nowhere. It is therefore inappropriate to suppress the reference to rationality and to confound objective knowledge with common belief.
2. Because the model posited by Samarqandī can be classified in sociological terms as *internalist*, it does not require that one take into account the role that other factors—political, economic, ideological and so forth—may play in the controversy.
3. As a consequence, it becomes possible to demonstrate the success or failure of the party to a scientific controversy by following the thread of the debate, whether there are external contributing factors (a frequent case in technological controversies) or not (the case of scientific controversies *stricto sensu*).

4. A Contribution to the Analysis of the Settlement of Controversies

Samarqandī's list of the signs of defeat makes a valuable contribution to the analysis of the settlement of controversies (see Engelhardt and Caplan 1987). In effect, one must not construct an external typology of the different ways that a controversy might end, but detect *in the exercise of the debate itself* the weaknesses that led to the victory or defeat of one of the disputants. Therefore, the settlement of a controversy should be studied solely on the basis of the rational practices of the debate that does not neglect its inner sociological dimension.

Samarqandī devised a powerful theory of scientific controversies that is free of the weakenesses and deficiencies of the modern relativist and constructivist epistemology. He laid the groundwork for a sociology of rationality inspired by a juridical model that does not deny the distinction between the true and the false and recommends the study of the settlement of controversies through the inner signs of defeat. The sociology of the sciences could learn a great deal from his model.

Notes

1. This text first appeared in French as "Al-Samarqandī. Un précurseur de l'analyse des controverses scientifiques," *Al-Mukhatabat* 7 (2013): 8–25.
2. This thesis is sustained by Collins (1981), Lynch (1985), Latour and Woolgar (1988), MacKenzie (1993), Knorr-Cetina (1995) and Restivo (2005) among others.
3. Even if the theory of the actor-network sometimes tries to deny this in the name of a "generalized principle of symmetry," the assumption that knowledge depends on social factors is present in case studies, as when Latour (1989) writes with regard to the controversy over spontaneous generation: "Pouchet rejects the commissions because he accuses them of ideological and political bias." See Chapter 2.
4. His known manuscripts are: 1. *Risāla al-samarqandiyya fī adāb al-bahth*, Istanbul, Süleymaniye kütüphanesi, MS. Ayasofya 4437; Süleymaniye, MS. Aşir Efendi 467; Paris, BnF, MS Arab 4378; Berlin, Staatsbibliothek, Mq. 119 [*henceforth referred to as Risāla]. 2. *Qustās al-afkār*, Istanbul, Topkapı, MS. Ahmet III 3399 [*henceforth *Qustās*]. — 3. *Sharh al-muqaddima al-burhāniyya*, Dublin, MS. Chester Beatty 4396; Cairo, Dār al-kutub al-misriyya, MS. Fiqh Taimūr 438 [*henceforth *Sharh*]. 4. *Mu'taqadāt*, Medina, 'Arif Hikmat Library, MS. Majāmī' 206 [*henceforth *Mu' taqadāt*]. It is beyond the scope of this chapter to address the philological aspect of Samarqandī's work; unless otherwise stated, the citations from the text of *Risāla* have been drawn from the transcription of Karabela (2010) and the translation of Miller (1984).
5. The same questions would be treated in the Latin world beginning in the thirteenth century in works on the *ars obligatoria*. The reference text for this tradition is Paulus Venetus, *Logica Magna* II, 8: *Tractatus de obligationibus* (ca. 1396–99) (Venetus 1988). We may also cite the various *Obligationes* written by William of Sherwood (1190–1249), Walter Burley (ca. 1274–1344), and Richard Swineshead (fl. ca. 1340–1355) (Spade 1977, Spade and Stump 1983, Stump 1985). Rebaptised *ars disputandi* in the sixteenth century, this current of thought would be integrated with the works of Josse Clichtove, *De artium scientiarumque divisione introductio* (1500), who was a pupil of Jacobus Faber Stapulensis (himself the author of *In terminorum recognitionem introductio*, 1504; and the works of Joannes Eckius, *Elementarius dialectica* (1517), Augustinus Hunaeus, *Erotemata de disputationes instituenda* (1558), Jacobus Reneccius, *Artificium disputandi* (1619), Conradus Horneius, *Liber de processu disputandi* (1633), and Abraham Calovius, *Tractatus novus de methodo disputandi* (1637) (Angelelli 1970; Felipe 1991). This corpus would profoundly influence the practice of the debate in classical Europe, most notably through the mediation of such scholars as Leibniz, who studied and wrote on the art of controversies (Dascal 2006).

Did the Arab theory of *adāb al-bahth* directly or indirectly influence the development of the *ars disputandi?* There are in fact more similarities than differences between the Arab and Latin traditions; according to Karabela (2010, 235–236), there are nine similarities as opposed to three differences. What is more, the differences are slight: (1) The Latin treatises on *ars disputandi* always include an account of the history of the dialectic while Arab treatises do not; (2) In the tradition of *adāb al-bahth* the disputants must enumerate all of the opinions on which they based their case, which was not a requirement for Latin authors; (3) Controversies in the Latin world were conducted before a presiding figure (*praeses, arbiter, moderator*) who could, when necessary, notify the parties when a formal error has been committed by one of the disputants; in the Arab world the debate was conducted between two or more parties, generally before an audience, for example the *majālis* (Stroumsa 1999). Apparently these similarities cannot be ascribed to the work of al-Samarqandī, because none of his texts were translated into Latin. European authors instead relied on other sources, principally Ibn Sīnā, *Kitāb al-Shifā' (Sufficientia)*, a text that was known to William of Sherwood and Paulus Venetus, and Ibn Rushd, *Talkhīṣ kitāb al-jadal,* which was published together with a Latin translation of Aristotle's *Organon* in an edition that was familiar, most notably, to Burley.

6.   A further clarification was proposed by a Karaite, al-Qirqisānī, who divided the questions into two groups. During the first and second stages of the debate the *opponens* obtains information from the *respondens*, while during the second and third stages the *opponens* attempts to refute the thesis of the *respondens* (Vajda 1963; Miller 1984, 21–22).

7.   In particular, this theory presents the same quirks as any pragmatic or linguistic theory of science that tends to favor the consistency theory of truth over the correspondence theory of truth. As a result this theory overestimates the possibility of arriving at the truth through discussion and underestimates the importance of referring to the external world in its analysis.

6

The SSK in the Name of Prestigious Ancestors: Duhem, Quine and Wittgenstein

The perennial difficulty encountered by SSK[1] in establishing the dependence of scientific knowledge on the social context in which it was produced can often be put down to an insufficient clarification of the underlying concepts,[2] which allows some observers to read systematic and necessary relations where others only perceive personal and contingent connections.

This would appear to have been the case, for example, when Norton inferred the continuity between eugenics and statistics from a detailed study by Karl Pearson (1978). It is generally acknowledged that Pearson's contribution to statistical methods was closely linked to his medical doctrine, because statistics can indeed contribute to furthering the study of genetic selection.

Nevertheless, the role of eugenics in the development of statistics is postulated rather than demonstrated. In order to make the hypothesis credible, we need to do more than find *one* eugenistic statistician; we must prove—at least over a well-defined period—that the four-cell eugenics-statistics contingency table contains more people in the eugenist-statistician cell and in the non-eugenist, non-statistician cell than in the other two cells. We first have to demonstrate that a *significant portion* of statisticians were in favor of eugenics, and already Bayes, Bernoulli, Condorcet, Laplace, Gauss, and Kolmogorov must be excluded. Vice versa, we also need to demonstrate that a *significant portion* of eugenists contributed to the advancement of statistics, but this was not true of Charles Darwin, August Weismann, Georges Vacher de Lapouge, Alexis Carrel, and Charles Richet. Therefore, as we extend our investigation into the two scientific communities of statisticians and eugenists, the personal correlation argued for in the case of Pearson disappears. Let us suppose now that this correlation can be testified to for an entire generation of scholars: should we speak in terms of "eugenic statistics"? Nothing could be less certain because we have good reason to use the terms "normal distribution" and "standard error," just as we use the terms "zero" and "compass" rather than "*Indian* zero" and "*Chinese* compass," because such terms presuppose universal knowledge that can be emancipated from the social context of discovery.

It is not rare to see relativist sociology, when involved in such an argument, justifying its theories on the basis of concepts drawn from the philosophy of science. This is a way of extending the debate. The aim of the present chapter is to propose a rereading of the continual exchange between SSK and epistemology, and to attempt to understand what SSK seeks to borrow from philosophical texts, with what objective, and to what extent it is within its rights to do so.

Let us recall the relativist tenets:

R1   The objects of the natural world that scientific statements are related to are nothing other than "textual constructions" (Woolgar, Latour).

R2   The natural world plays a negligible role in the construction of scientific statements (Collins).

R3   The social context, local as well as global, plays a decisive role in the construction of scientific statements (Mulkay).

R4   Scientific knowledge is "conventional" (Bloor) and its reasoning is built on "informal social negotiation" (Mulkay).

The work of the philosophers Duhem, Quine, and Wittgenstein has been crucial to the supporters of relativism in the development of their own ideas. This emerges clearly in the writings of Collins (1974), Bloor (1973, 1976, 1983), Barnes (1977, 1983), Cartwright (1983), Latour (1984, 1987), Callon and Latour (1991), Shapin and Schaffer (1993), Fourez (1996) and others. Let us take just one example of the connection that has often been made between the sociology and the philosophy of science. In his textbook on the sociology of science, Dominique Vinck (1995) suggests that the philosophers of science, by a radical internal shakeup, "opened the door to a sociological analysis" of the content of science that had remained suspended in an embryonic state in the sociology of Merton (1937, 1938, 1973). But the question remains: To what extent can references to these philosophers be used to support the SSK program? This is a crucial question for the future development of the sociology of knowledge.

1. Pierre Duhem

Recently there has been renewed interest in the work of the physicist and philosopher of science Pierre Duhem (1861–1916), whose books have been reissued (1980, 1985, 1991) with detailed analysis and commentary (Vuillemin 1981; Ackermann 1985; Crowe 1990; Brenner 1990; Hacking 1992; Stoffel 2002; Darling 2003).

1. Epistemic Holism

His ideas on "epistemic holism" have drawn the attention of sociologists,[3] in particular passages such as this one: "The physicist can never subject an isolated hypothesis to experimental testing,[4] but only a whole group of hypotheses; when the experiment is in disagreement with his predictions, what he learns is that at least one hypothesis constituting this group is unacceptable and ought to be modified; but the experiment does not designate which one should be changed" (1991, 187). From this Duhem derived a critique of the inductive method and a profound revision of Bacon's concept of the "crucial experiment." Holism thus leads to a weakening of the notion that scientific content can be determined by means of experiments on the natural world. *Prima facie*, this would seem to lend weight to the relativist position and to the idea of the dependence of knowledge on social beliefs. This connection, however, comes up against two main arguments.

First argument. In his texts on holism, Duhem never proposes the chain of connection R2-R3 imagined by SSK. He is categorically opposed to any such attempt.

Primo, Duhem confines his analysis to the science of physics (1981: xv); on many occasions when other disciplines, such as mathematics or physiology, might have been introduced they do not receive his attention. Therefore, we cannot jump to any conclusions regarding scientific activity in general based on his work.

Secundo, Duhem (1991) firmly denies the existence of a metaphysical (or religious) foundation to this science: "Physics proceeds by an autonomous method absolutely independent of any metaphysical opinion" (274). He shows, for instance, that the contributions of Descartes, Huygens, and Fresnel to optics were not deduced from the principles regarding the nature of light that these scientists firmly adhered to, but from the experimental results that they obtained. Duhem developed this point in 1905 in response to Abel Rey, who criticized him for inappropriately introducing Catholic creeds into his study of the methodology of physics. In this text Duhem (1991) reaffirms his belief in the separation of physical and metaphysical questions: "Between two judgments not having the same terms, there can be neither agreement nor disagreement" (283).

Tertio, Duhem (1991) does not admit the transfer of certain hypotheses from "common sense" to physics. He explains himself by showing that "The fund of common sense is not a treasure buried in the soil [. . .]; it is the capital of an enormous and prodigiously active association formed by the union of human minds." This is the only occasion on which he utilizes a sociological term in his text, before concluding that, if a physicist believes himself to be using a hypothesis based on common sense, "[. . .] he will simply have to withdraw from the fund of common-sense knowledge the money that theoretical science had itself deposited in that treasury" (261).

These three reasons, expressed without equivocation by Duhem, invalidate the relativist principle that society plays a major role in the construction of scientific statements (R3). Nothing could be farther removed from his view.

Second argument. Duhem's epistemic holism paves the way for a question that Jules Vuillemin and Anastasios Brenner have echoed. When we say that the hypotheses of a theory make up a coherent whole, what should we understand by the concept of "whole"? Are we limiting it to a specific part of a scientific subject, to a specialized field, to physics in general, or to the whole of human knowledge? From his writings one gains the impression that Duhem restricted his epistemic holism to a single discipline (physics), and that it resembles the holism of physical bodies subjected to gravity. Indeed, in order to test certain hypotheses in the field of mechanics, Duhem (1991) writes: "[. . .] it would be necessary for isolated systems to exist. Now, these systems do not exist; *the only isolated system is the whole Universe*" (295, italics mine). However, a systematic search for passages that mention the word "whole" shows Duhem's hesitancy. Sometimes he appears to be referring to the whole system of "physical theory" (220), whereas at other times he seems only to mean the identifiable and dissoluble *elements* that form part of a scientific statement (292). Vuillemin and Brenner have interpreted this hesitation in opposite ways.

Vuillemin (1986) attempted to limit the impact of Duhem's epistemic holism by proposing the idea of a "compartmentalisation" of the sciences. He writes: "I understand by 'compartments' the existence of autonomous and almost closed systems that closely resemble the ideal of a system independent of any external intrusion. The history of taxonomy, astronomy and dynamics shows that science has been made possible because some of the compartments were frequent and basic enough [. . .] to lend themselves easily to theoretical reconstruction" (19). Therefore, Vuillemin's restriction consists in declaring that a theory should be tested on the basis of the propositions in the field to which the theory belongs. Let T be a theory belonging to compartment C of a science S, and let p be a proposition of T, p' a proposition of C – T, and p'' a proposition belonging to S – C. The relationship Fpp' (p' grounds p) is false if there exists a p'' such that Fpp''. And yet we can find many examples of this type. Let us take the example provided by Duhem (1991), which consists in the calculation of the position of the sun (169–171). This involves several hypotheses: geometrical (based on the reduction of the solar globe to a sphere and the reduction of the center of gravity to a geometric center), optical (the constancy of the speed of light, the law of atmospheric refraction), temporal (knowledge of solar time and sidereal time), geographical (the determination of latitude and longitude), or mathematical (algebra, sexagesimal arithmetic and trigonometry). As these hypotheses are external to compartment C of mechanics, they contradict the limitation of holism proposed by Vuillemin (1986).

Does this example justify the interpretation of Brenner (1990) that the testing of a theory implies addressing *every* pertinent proposition of physics? Assuming the same notational conventions (let T be a theory of compartment C of a science S, p a proposition of T, p' a proposition of C – T, and p'' a proposition of S – C), then Brenner's reading corresponds to Fpp''. This relationship is false if a p'' exists, so that $\neg Fpp''$. Once again, in the same example it is clear that if external hypotheses are necessary to determine the position of the sun, no notion of relativity, electrostatics, or thermodynamics will be useful for this calculation. Many compartments of physics do not contribute at all to the setting of T. Thus, the interpretation envisioned by Brenner (1990) is also questionable.

The different readings of Duhem's epistemic holism proposed by Vuillemin and Brenner raise difficulties that call for a fresh interpretation. If we return to Duhem's example of the calculation of the position of the sun, it is clear that, given the compartment of mechanics C, not every proposition of C grounds T, and the propositions that actually ground T do not necessarily belong to C. Equally, not every proposition of S grounds T.

Perhaps Duhem's notion of "holism" is that *certain theoretical elements* which come into the picture of a physical theory belong to the whole field of a science, and not that *every* theoretical element of this science contributes to the construction of a specific theory. Vuillemin's compartmentalism can be retained if we introduce—with due caution—the negative statement: "Some compartments of a science *do not* contain any proposition founding a theory." In this case Duhem's epistemic holism would not constitute a holism *stricto sensu*. Indeed, it would lose its nature as an organic whole and express no more than the need to ground a theory on something *external*.

Whichever interpretation one accepts, it appears that none of them can reconcile Duhem's ideas with the principles of relativism. In particular, R3 is diametrically opposed to the Duhemian view, which would never admit a continuity between social beliefs and physical theory.

2. Underdetermination of Theory

Duhem's (1991) thesis concerning the underdetermination of theory (which would be developed by Quine)—that several rival theories may pass the same experimental tests—is also much quoted. "An infinity of different theoretical facts may be taken for the translation of the same practical fact [. . .] But an infinity of other expressions might just as well have satisfied these requirements" (18). The same physical phenomenon can be described by several incompatible theories. In *Sôzein ta phainomena*, Duhem (1969) proposes—in keeping with Andreas Osiander (1498–1522) and Robert Bellarmine (1542–1621)— that "the hypotheses of physics are mere mathematical contrivances devised for the purpose of saving the phenomena" (117). This passage contains unmistakeable overtones of relativism.[5] But does it constitute a sufficient guarantee of the SSK stance—in particular, the principles R2 (the world plays a negligible role in the construction of scientific statements) and R4 (every theory is conventional and based on a fragile consensus)? New and well-founded objections have been raised against the idea that these principles can be deduced from Duhem's texts.

First argument. Duhem (1991) admits the arbitrariness of theories in physics, but states that this arbitrariness is confined to its stage of development. Admittedly, he writes: "A physical theory is free not to take account of experimental facts" (206). But this sentence should be read with caution, for Duhem hastens to add: "This is no longer the case when *the theory has reached its complete development*" and when it is subjected to experimental tests (that will decide whether or not it has to be rejected). Elsewhere Duhem explains that an array of rival theories may be reduced to just one, even *before* being tested, depending on the usefulness of the deductions that can be obtained from them (137). In addition, the fact that two rival theories are in a position to explain the same set of facts does not prevent us from preferring one to the other in a rational choice based on the logical aspects of the two theories[6] (for example, one theory may require the acceptance of some *ad hoc* and inconsistent hypotheses, whereas the other may offer arguments of self-evident clarity). Duhem (1981) cites the example of Biot, who refused to support Newton's corpuscular theory of light once it had been contradicted by Foucault's experiments (218).

This development shows that we cannot deduce the principles R2 and R4 of relativism from Duhem's texts, without running the risk of a forced reading. The misunderstanding stems from reasoning *pars pro toto*. The fact that the development of scientific hypotheses can be weakened by arbitrariness does not mean that experimental tests and the deduction of consequences will necessarily be so vitiated. It must be remembered that in Duhem's view such operations are designed to limit the arbitrariness of the initial hypotheses.

Second argument. Regarding the theory of the conventional nature of knowledge often ascribed to him, Duhem does not adopt an uncompromising position, for he explicitly subscribes to both the theory of consistency (formal rationality) and the theory of correspondence (objective rationality).

Primo, the physicist is free to construct any theory he likes so long as it conforms to the rules of logical consistency. Duhem (1991) often returns to this aspect of theories in physics, writing: "With these materials, theory builds a logical structure; in drawing the plan of this structure it is hence bound to respect scrupulously the laws that logic imposes on all deductive reasoning" (205). The philosopher then applies these characteristics to the hypotheses: "In the first place, a hypothesis shall not be a self-contradictory proposition, for the physicist does not intend to utter nonsense. In the second place, the different hypotheses which are to support physics shall not contradict one another" (220). The arbitrariness of a theory is thus conditioned in two ways: by time (its stage of development) and by an external norm (logical consistency).

Secundo, Duhem (1991) shows in fact that the limits to arbitrariness are not two-fold but three-fold. A hypothesis is indeed subject to the requirement that it correspond with the real, which Duhem calls "representation" or "likeness" (286). When discussing the experimental consequences that derive from hypotheses, he writes: "These judgements are compared with the experimental laws which the theory is intended to represent. If they agree with these laws [. . .] the theory has attained its goal, and it is said to be a good theory; if not, it is a bad theory, and it must be modified or rejected [. . .]. *Agreement with experiment is the sole criterion of truth for a physical theory*" (21). However, this adequacy is never perfect, for it is based on a translation of natural phenomena into a symbolic language that surreptitiously introduces approximations. Thus, it becomes clear that "every physical law is an approximate law" (171). This is why any physical law is temporary and revisable; not only because of the freedom of construction that theoretical activity presupposes, but also because of advances in the methods of measurement and observation (172). The increasing accuracy of physical measurements today requires the constant revision of scientific hypotheses, in a process that leads to increasing agreement between theory and phenomena (204). Duhem cites an example to illustrate this progress: "When we see [. . .] the vast domain of optics, hitherto encumbered by so many details in so confused a way, become ordered and organized, it is impossible for us to believe that this order and this organization are not the reflected image of a real order and organization [. . .] The more complete it (i.e., the physical theory) becomes, the more we apprehend that the logical order in which theory orders experimental laws is the reflection of an ontological order" (26).

Duhem therefore does not simply recognize the existence of the real; he admits that, by running trials, physics manages to build up a set of statements that corresponds to it with ever increasing accuracy, and the final goal of the physical theory is to become a "natural classification." This unequivocal position is in clear conflict with: principle R1, that the world does not exist outside textual constructions; principle R3, which minimizes the function of the natural world in the construction of theories; and principle R4, which expounds the conventional nature—in a non-Duhemian sense—of scientific theories. Whatever we may think of Duhem's idea of physics, we should recognize that it does not lend support to the relativist position of SSK. In Duhem's view the preference for one theory over another is not determined by social beliefs, but by considerations of adequacy, simplicity, and mathematical elegance.

2. Willard Quine

The works of the American philosopher and logician Willard van Orman Quine (1908–2000) offer an interesting expression of the thesis of the "underdetermination of theory" that is often utilized by sociologists to support their position. Springing from the analytic tradition and Quine's own critique of logical empiricism, this development was marked by subtle differences that partially invalidate the label "Duhem–Quine thesis" applied by certain scholars (see Quine 1963, 41; 1977, 17).

1. Epistemic Holism

Quine's (1974) holism returns to the main argument of *La Théorie physique*; indeed we note an almost identical position on certain points: "Our statements about external reality face the tribunal of sense experience not individually but as a corporated body" (2). According to Boyer (1978), Quine developed the argument even further: 1) he converts Duhem's neutrality into an "ontological engagement"; 2) he abandons the dualism of scientific and commonsense knowledge in favor of the notion that there is a continuity between them; 3) his holism concerns not only the interpretation of experiments, but also the meaning[7] of theory (semantic holism), thus inducing Lakatos to distinguish between a weak version (Duhem) and a strong version (Quine) of the "Duhem–Quine thesis" (Lakatos 1980, 97).

First argument. Let us start with epistemic holism, which in Quine's (1960) clearest elucidation is defined as: "Experiences call for changing a theory, but do not indicate just where and how" (58). Quine nevertheless spontaneously limits his holistic thesis,[8] beginning with the assumption that we are not compelled to revise all the propositions on which a theory is based, but only those most directly tied by relations of "affinity" or "germaneness" to the object of investigation. He (1953) writes: "In this relation of 'germaneness' I envisage nothing more than a loose association reflecting the relative likelihood, in practice, of our choosing one statement rather than another for revision in the event of a recalcitrant experience" (43). Two classes of statements enjoy a greater degree of stability: observational statements and logico-mathematical statements. Quine (1974) himself explains the immunity that he grants to the second class of statement:[9] "The more fundamental a law is to our conceptual scheme, the less likely we are to choose it for revision [. . .]. Vast domains of law can easily be held immune to revision on principle [. . .]. Mathematics and logic, central as they are to the conceptual scheme, tend to be accorded such immunity, in view of our conservative preference for revisions which disturb the system least" (2). The idea of a selective revision of hypotheses—sometimes referred to as the rule of "least action" (2)—can be connected with Vuillemin and either his compartmentalism or his restriction, as suggested above. But in both cases it overtly challenges the tenets R1 and R3 of relativism.

Second argument. A serious incompatibility seems to exist between Quine's holistic view and the conventionalism that we have identified as principle R4 of relativism. Indeed, the intent of epistemic holism is to retain the criterion of verification, otherwise we could not even say that scientific statements compare the results of experimental tests (as an organized body), because no verification procedure exists that allows us to compare them with reality. Therefore, holism and verificationism are interdependent. If the meaning of a theory is based on procedural rules, then it does not have to be founded on *a priori* conventions. This result has unexpected consequences. Because Quine (1980) questioned the existence of a gap between synthetic and analytic statements, the previous conclusion applies both to the experimental sciences and to logical-mathematical domain, where the notion of convention is similarly suspended. This allows Michael Seymour (2000) to write: "Logic no longer assumes the guise of an ensemble of conventional rules, but of an effective inferential practice, and the rules of calculation only exist to shape this practice" (136), a statement that invalidates the deduction of principle R4 from Quine's texts.

Third argument. Finally, Quine detected an incompatibility between semantic holism and the thesis of the underdetermination of theory (as well as its corollary, the indeterminacy of translation). Indeed, semantic holism is closely connected with the verificationism of the 1930s. The Vienna Circle linked the meaning of a statement to its truth-conditions (or, in what comes to the same thing, to its verification procedure). It follows that if two theoretical statements have the same truth-conditions they are *synonymous* and disprove the thesis of the underdetermination of theory (Gochet 1978, 40). As Quine was seeking to preserve this thesis, he moved away from holism towards a form of "semantic atomism." Along these lines he (1960) writes: "Translation proceeds little by little and sentences are thought of as conveying meaning severally" (71). In a way Quine found himself constrained either to abandon the thesis of the underdetermination of theories in favor of holism (which clashes with the arguments above) or to abandon holism in favor of the thesis of underdetermination, as we shall now consider.

2. Underdetermination of Theory

Quine's (1960) version of the underdetermination of theory by data implies that various rival theories are in a position to pass experimental tests *ex aequo*. "In general the simplest possible theory for a given purpose is not necessarily unique [. . .]. Scientific method is the way to truth, but it affords even in principle no unique definition of truth" (21). This position refutes the idea admitted by Duhem that science may approach the truth asymptotically. What we seem to have here is a point that is congruent with the relativist principle R4, that every truth is of a conventional nature and lends itself to negotiation. This reading, however, is subject to debate.

First argument. The difficulty in deciding between various theories stems from an equivalence between rival systems—from the point of view of experiment or logic or scientific interest. In such cases scientists may agree to adopt one theory, but there is very little chance that they will do so if the theories present *exactly* the same degree of correctness or usefulness. The history of science abounds with examples of rival theories, both of which are upheld for a long period before a new experiment or logical revision comes along to prove the superiority of one over the other. Consider, for instance, the lengthy debate (1815–1919) regarding the atomic weight of chlorine (discussed in Chapter 5). On a smaller scale, every scientific controversy provides confirmation of this. The postulate that one should be able to reduce any set of valid truths conventionally to a single, unique truth is wrong, because no rule or behavioral norm will allow a scientist to choose between several theories that are *exactly* equivalent. Therefore, either there is no choice because the theories are strictly equivalent (the conventional *choice* is groundless) or there is a choice that can be made on the basis of scientific criteria (the *conventional* choice is groundless). The conversion of the thesis of the underdetermination of theory into the thesis of the determination of theory by social factors (principle R3) does not take any regular form, which is probably the reason why this conversion is never made explicit.

Second argument. If we examine Quine's work, it does not take long to see how the thesis of the underdetermination of theory correlates with relativism. He (1960) asks: "Have we now lowered our sights so far as to settle for a relativistic doctrine of truth? [. . .] Not so" (22). Such a negative answer requires an explanation. Let us take the example of two rival theories stemming from a rewording of the laws of mechanics. The relationship that links force, mass, and acceleration admits of two equivalent wordings, in the form of either a differential equation or a discrete equation. Some have supposed this rewording to constitute a case in which two theories have exactly the same empirical content, but are logically incompatible—in the first equation, time is continuous, while in the second it is discontinuous.

Although he is clear about his goals, this example is questionable.

Primo, it does not fit Quine's purpose. Quine (1960) writes: "The indeterminacy that I mean is more radical. It is that rival systems of analytical hypotheses can conform to all speech dispositions [. . .] and yet dictate, in countless cases, utterly disparate translations" (66). A discrete equation may be transformed into a differential equation and vice versa. Therefore, they represent two "paraphrases."

Secundo, if we assume that they are not interchangeable paraphrases, their difference in relation to the model of time removes any strict equivalence between them. A choice is then possible between the two translations, depending on a norm of correspondence (the model of continuous time showing better agreement with experience). It is difficult to find an indisputable case of two statements having exactly the same empirical content.

Third argument. Let us admit, for argument's sake, that Quine comes close to adopting a relativist idea of truth, as some of the proponents of the SSK appear to think. It is true that Quine (1993, 116–117) admits a deflationist theory of truth, such that:

"*p*" is true if *p*

Thus, the statement "the table is round" is true if, and only if, the table is round. This disquotation raises at least two problems. *First*, as Engel (1998) observes, it seems difficult "to avoid reintroducing here our intuitive notions about correspondence and consistency" (38). For instance, are we entitled to disquote the sentence in all cases? A detailed examination shows, on the contrary, that we can disquote it only if there is a way of subjecting the roundness of the table to verification. Otherwise, it seems impossible to know what "the table is round" means. In this context, the disquotation theory of truth is reduced to nothing more and nothing less than the classical correspondence thesis. *Second*, the disquotation theory conflicts with the pragmatist definition of meaning that Quine adheres to. It stipulates that the meaning of a statement is determined by its truth-conditions, whereas according to the disquotation theory the examination of the possible truth of a statement presupposes its meaning. It requires either that one abandon verificationism—as Quine refused to do in other circumstances—or that one revise the theory of truth in favor of the correspondence thesis, as would appear to be possible.[10] But the fact remains that both possibilities weaken the principles of relativism R1 and R2.

Each conclusion requires that we give up one or more points of the relativist program; therefore, it is not possible to endorse Quine's view as a whole without—as a consequence—refuting the fundamental principles of SSK.

3. Ludwig Wittgenstein

The texts of Ludwig Wittgenstein (1889–1951) have often been cited by relativists in the field of SSK, in particular by David Bloor (1973, 1983b). It is true that Wittgenstein was more interested in investigating the play of language than the nature of science. It is true that Wittgenstein defended an antirealist position that is hardly (or not at all) compatible with the practice of science. It is true that the texts of Wittgenstein, more than those of any other modern philosopher, have been subjected to endless interpretation; this can be observed down to his preference for "uses" rather than "definitions." However, these traits are not sufficient to justify viewing Wittgenstein as a *legitimate* precursor to sociological relativism. As with Duhem and Quine, it is necessary to test soberly the solidity of the link between the texts of Wittgenstein and SSK.

It is not the early work of Wittgenstein that has attracted the attention of the proponents of SSK, but the speculations that he embarked upon in Cambridge beginning in 1929 and that led to the writing of the *Philosophical Investigations*. The question still remains open as to whether it is correct to distinguish between an "early" and a "late" Wittgenstein. His two groups of texts—the first produced while he was still in Austria and culminating in 1921 with his *Tractatus Logico-Philosophicus*, and the second produced between 1936 and 1949, after he had found refuge in Cambridge—unquestionably reflect an evolution in his thought but they cannot be divorced from one another, as certain commentaries have implied.

Scholars who have studied the Viennese philosopher's writings have underlined his unique, formulaic mode of expression, which lend his statements their aphoristic quality. Such were the strength of his ideas and the literary purity of many of his written passages that Shwayder (1969) has described their effect as being like "flashes of lightning" (66). All the same, the term "aphorism" presents a danger as it implies that each paragraph by Wittgenstein can be read autonomously and his ideas extracted arbitrarily. *Apophthegma*—the Greek term for "memorable words"—would be more appropriate, because it places responsibility for the excision process on the commentator.

Although Wittgenstein's two seminal works—the *Tractatus* and *Philosophical Investigations*—both take the form of structured successions of paragraphs, the approach to his texts that consists in extracting passages in order to synthesize a view is not in keeping with the author's own intentions. Wittgenstein (1961) refuses to condense his thoughts (111). The temptation to construct such syntheses may be difficult to resist, but the process can lead to as many different results as there are possible choices of passages to excise. There is a visible order to the *Philosophical Investigations* that follows the hierarchical organisation of the *Tractatus*. In general, the extraction of a remark includes the deletion of the numeration assigned by the author, which is seldom retained as a reference[11] even though this ordering number indicates precisely which propositions the remark should be linked to. The *Tractatus* is not a book or an album. It is a *tree*, and the reader is supposed to make a clear distinction between two types of progression: a "vertical" reading that links coextensive propositions (e.g., 3.11, 3.12, 3.13, 3.14); and a "horizontal" reading which follows the movement of explication of a given proposition (e.g., 2, 2.1, 2.13, 2.131) (Granger 1969, 1990, Figure 5). Granger then summarizes the chain of basic propositions in these terms: "The world is all that is the case; A logical picture of facts [a proposition] is a thought; Truth-possibilities of elementary propositions are the conditions of the truth and falsity of propositions; What we cannot speak about we must pass over in silence" (see Granger 1969, 22–25).

As Wittgenstein himself admits, the moderation in position that we note between his *Tractatus* and his *Philosophical Investigations* marks a change in his treatment of the problem of meaning. But Wittgenstein's suggestion that the two groups of apophthegms be taken together also testifies to a certain continuity between them (as, for example, in his endless interrogation of the concepts of activity and use). Clearly, as far as Wittgenstein is concerned the *Tractatus* and the *Philosophical Investigations* are interdependent. When we read passages from the latter, we can find continually their roots in the former. Thus, paragraph 282 is an extension of the *Tractatus'* fourth horizontal chain (4, 4.4, 4.46, 4.461, 4.4611).

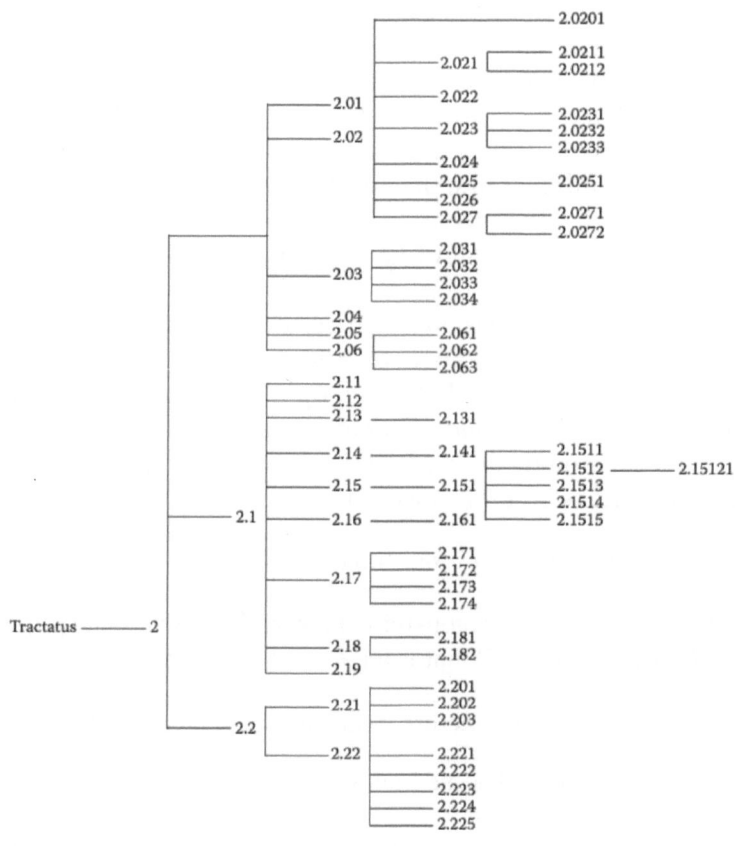

Figure 5. The tree of the *Tractatus*, proposition 2.

This analysis of the method of reading required by his apophthegms allows us to address the passages in Wittgenstein on which sociological relativism claims to be based.

1. *The Conventional Nature of Knowledge*

Let us start with the conventional nature of knowledge. Bloor (1983a, 34), Barnes (1983, 33), and Shapin and Schaffer (1993, 152) have viewed Wittgenstein's writings as a source of support for their program of studying the *social foundations* of (conventional) scientific knowledge. In so doing, sociological relativists ignore the fact that Wittgenstein himself posed a very similar question: "Are the propositions of mathematics [merely] anthropological propositions saying how we men infer and calculate? Is a statute book a work of anthropology telling us how the people of this nation deal with a thief? [. . .] Well, the judge *does not use* the statute book as a manual of anthropology" (1956, 67). Mathematical conventions are not anthropological objects. But let us suppose all the same that Wittgenstein remained silent on the *extrasocial* nature of conventions, and continue our examination of the relativist interpretation of his thinking. The passage that fits their view best is:

4.002. The tacit conventions on which the understanding of everyday language depends are enormously complicated.

Taken on its own, it is true that this apophthegma bears some relativist overtones. But as we made clear above, an aphoristic reading of Wittgenstein's statements would be mistaken. His proposition must first be linked to its source-proposition:

4. A thought is a proposition with a sense.

And what is thought? Wittgenstein answers: "3. The logical picture of the facts is the thought." What is a picture? Wittgenstein writes:

2.1511. *That* is how a picture is attached to reality; it reaches right out to it.
2.1512. [The picture] is laid against reality like a measure.

Thinking thus requires a correspondence between thought and the world, as 2.1511 and 2.1512 imply. There is no trace of nihilism or skepticism here. The exercise of thinking does not mean that man is condemned to the manipulation of arbitrary pictures (because the tacit conventions of understanding the world would be unattainable), nor that the world does not exist: "1. The world is all that is the case."

In 4.002, Wittgenstein refers only to the difficulty of describing the functioning of *natural language*, which is not to be confused with *formal language* (this is precisely the problem that will encourage him to set about describing natural language from 1929 onward). Wittgenstein makes a distinction between philosophy, which uses natural language, and science, that may employ a formal language. This demarcation is established by the following paragraphs:

> 4.11. The totality of true propositions is the whole of natural science (or the whole corpus of the natural sciences).
> 4.111. Philosophy is not one of the natural sciences.

Therefore, paragraph 4.002 cannot be pressed into service to support the SSK program, because Wittgenstein assumes the following: the scientist's goal is to describe the world, he achieves this end by constructing a logical picture, and he can determine whether or not this picture corresponds with the real. The relativist reading is not justified, insofar as it attaches no importance to the clear distinction that Wittgenstein makes between statements expressed in common language and scientific statements.

2. The Language-Games

Let us now focus on one of the most famous passages in the *Philosophical Investigations*, in which Wittgenstein introduces the notion of "language-games." According to Shapin and Schaffer (1989): "The term 'language-game' is meant to bring into prominence the fact that the speaking of language is part of an activity or of a form of life" (15). Thus, they suggest that scientific controversies should be studied as "language-games." Other sociologists, including Vinck (1995), retain that natural language lay at the very centre of Wittgenstein's thought, and that he "denies any preferential place for logic" (85). Let us examine the objections that can be raised against these interpretations.

First argument. Apart from any considerations regarding its form of expression, a language is structured by its function. The goal of a game is to play; the goal of science is to describe, as far as possible, natural phenomena. To draw a parallel is to admit either that the correspondence norm applies to the game or that it does not apply to science. The first hypothesis is quite unconvincing, for a game of chess cannot exist *before* the players know the rules of the game. Such rules do not give an account of preceding experimental data; their aim is to fix the *a priori* conditions for the possibility of a game that is never given in advance. The second hypothesis is no better. It implies that science is a game whose rules are laid down *a priori*, and these rules fix the appearance of natural phenomena. An experiment to determine the presence of gravity, for instance, will not follow but *precede* the description of natural phenomenon. Wittgenstein—who sometimes expressed doubts himself about the analogy between games and language—would probably not have admitted this annexation of science by games. Science cannot be considered a game; the natures of the two are irreconcilable.

Second argument. Let us suppose now (and we are forced to make this assumption if we wish to continue) that Wittgenstein never spoke about the differences between mathematics, logic and natural language (see 1961, passim chain 4; 1965, 28). If this is the case, then we should wonder, together with certain other scholars, whether his reflections on language can lend weight to the SSK position. It should first be noted that, in Wittgenstein's view, the formula "language-game" does not imply the existence of a playful aspect to language, but the existence of *rules*, just as in games. This development follows paragraph 3.326 of the *Tractatus*, which bases the recognition of symbols on their use. In devising the term "language-game" Wittgenstein's objective was to clarify our use of words by applying the formula: "How do we use . . ." But what is this use? Wittgenstein (1961) suggests an analogy: "Let's say that the meaning of a piece is its role in the game. [. . .] The game must be determined only by rules!" (281–282). Furthermore, he (1983) always takes as his model games in which the rules are *well-defined, explicit, and compulsory*. "If a rule does not compel you, then you are not following any rule" (329). In addition, he (1965) declares language to be a human system of communication (81). When connecting these two developments in his thought, it is clear that we should interpret "language-game" to mean "explicit and compulsory rules of communication." If we do so however, it is probable that the advocates of SSK will be much less interested in the concept! Indeed, it would exclude any relativist interpretation. We may also add that whenever Wittgenstein (1961) admits the existence of "vagueness" in the rules, he does so only in the context of natural language, and carefully excludes the formal languages of science (162).

Third argument. Let us return now to the study of scientific conventions. In paragraph 199 of the *Philosophical Investigations,* Wittgenstein (1961) asks a question that invites a sociological reading of language-games. "Is what we call 'following a rule' something that it would be possible for only *one* person, only *once* in a lifetime, to do?" The answer is: "To follow a rule [. . . and] to play a game of chess, are *habits* (usages, institutions)" (202). Even if Wittgenstein limits his comment to natural language, it is clear that in this passage there is little difference between logic and ordinary language. What does the equivalence between uses, habits, customs and institutions mean? That sociology has something to tell us, not about the *rule,* but about the *obeying* of the rule.[12]

Does this imply that Bloor's (1991) declaration, "[. . .] scientific theories, methods and acceptable results are social conventions" (43) is in keeping with Wittgenstein's idea that no mental process is able to ensure that individual behavior will conform to the application of given rules? Let us assume that any act of obedience has a conventional dimension. There would still be no guarantee that "theories, methods and results" are identically subject to the obedience relationship. Do scientists obey theories as they obey the rules of calculation? Nothing could be less certain, since their work consists precisely in testing these theories on the basis of logical and experimental rules. So much so that the question of obedience does not apply to scientific knowledge, but only to the methods and the general norms that serve to test them. In other words, the ideas of convention and obedience apply in different ways to methods and scientific results.

The distinction proposed here between methods and results is one of major importance, because the question of the conventional nature of a method bears no relation to the question of the conventional nature of the results reached by that method. There is an absolute separation between the two. A good example is the process of multiplication. Many different algorithms for multiplying two numbers *a* and *b* were invented more or less independently: the Chinese approach (based on counting the number of points at the intersection of lines), the Persian method (with the deletion of intermediary results), the Indian method (wich retains the intermediary results), and the lattice method developed by the Arabs. Georges Ifrah (1981, 2:311–340) explained in full the algorithms developed by Naṣīr al-Dīn al-Ṭusī, Bhāskarāshārya and Brahmagupta . . . To be complete, any account of the procedures that can be used to multiply two numbers would have to include several finger calculus techniques as well (Pacioli 1494, 36v; Ifrah 1981, 1:149–153). The anonymous *Larte de labbacho* (1478, unpaginated), and the work of Luca Pacioli (1494, 25v–29r) and Giuseppe Maria Figatelli (1678, 6–12) provide evidence that many multiplication algorithms were being used in Italy, such as, inter alia, the chessboard *(scacchiera)*, the castle *(castelluccio)*, the lattice *(gelosia)*, the grid *(quadrilatero)*, the cross *(crocetta)*, the cup *(coppa o bicchiere)*, the pyramide *(piramide)* (Plate 7). These algorithms, which any person carrying out calculations obeys in a conventional manner—in keeping with

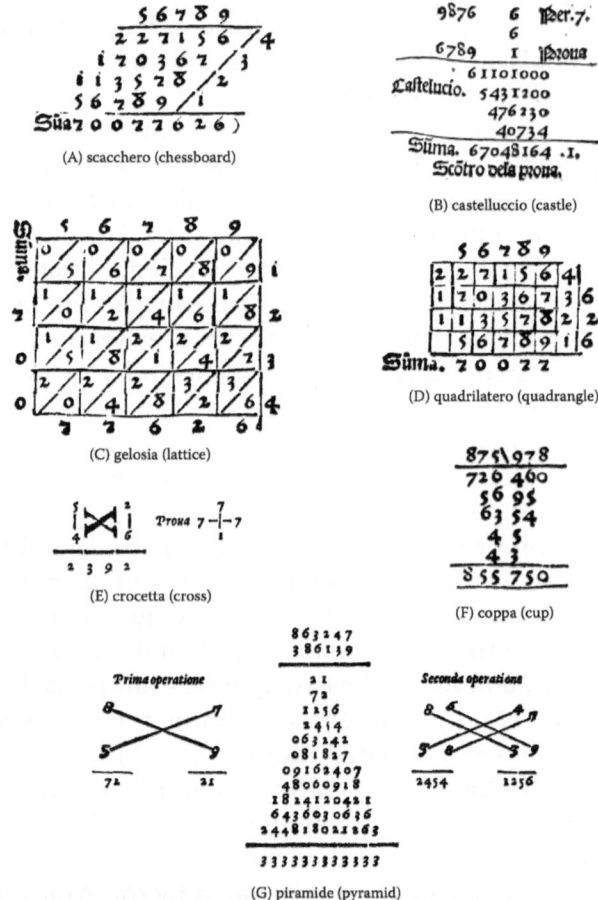

(A) scacchero (chessboard)

(B) castelluccio (castle)

(C) gelosia (lattice)

(D) quadrilatero (quadrangle)

(E) crocetta (cross)

(F) coppa (cup)

(G) piramide (pyramid)

Plate 7. Various multiplication algorithms, reproduced from: *Larte de labbacho*, 1478 (A, C, D); Pacioli's *Summa de aritmetica*, 1494 (B); and Figatelli, *Tratatto aritmetico*, 1678 (E, F, G).

Wittgenstein's mode—are very different from each other. However, there is no doubt that, once the *operans* and *operandum* are fixed, these *different* methods all yield the *same* results.

We can hardly see—even in the context of Wittgenstein's conventionalism—how theories and scientific results may proceed from social conventions. But it is still possible to justify Bloor's view by declaring that the conventional aspect of theories stems *indirectly* from the methods applied (because any theory relies on a set of conventional methods). We now need to decide whether or not methods are conventional in nature. Although this is often maintained, it forces us to confront a crucial paradox that was first pointed out by Lewis Carroll in 1895 in the form of a *modus ponens*. What must we do to infer the conclusion "*q*" from "*p*" and "*p* implies *q*"? From a strictly conventionalist standpoint, we can make such an inference only because of the conventional nature of the *modus ponens*: "if *p*, and *p* implies *q*, then *q*." But nothing guarantees that this rule will apply to the particular case we are focusing on. Therefore, we must admit by convention that: "if *p*, *p* implies *q*, if *p* and *p* implies *q*, then *q*." This opens the way for an infinite regression (Table 8).

A strict conventionalist point of view would then have to claim that the most elementary logical deduction requires an infinite number of premises. This result is clearly counterintuitive, because nobody can examine an infinite series of premises in a logical deduction. If *p*, and *p* implies *q*, we immediately deduce *q*, without resorting to any other premise. As Seymour (2000) writes: "It is hard to see how the conventionalist conception can appropriately solve Lewis Carroll's paradox" (130).

To summarize, if we rule out "aphoristic" readings that rely on extracting passages arbitrarily from texts, we observe that Wittgenstein's ideas—although relativist in nature—can only with difficulty be considered compatible with tenets R1 and R4 of the SSK.

Table 8. Lewis Carroll's paradox.

premises	conclusion
1. "*p*," "*p* implies *q*"	"*q*"
2. [1] + "*p* and *p* implies *q*, then *q*"	"*q*"
3. [2] + "*p*, *p* implies *q*, if *p* and *p* implies *q*, then *q*"	"*q*"
4. [3] + etc.	"*q*"

4. Conclusions

It is no small paradox that the texts of Duhem, Quine, and Wittgenstein so frequently cited by the SSK entirely fail to support the program that the SSK has attempted to base itself upon for nearly thirty years. A relativist *interpretation* of any text can always be made, but in such cases one becomes dependent on an "excising rule" that must at least be made explicit. It follows that references to these philosophers' works do not provide any guarantee of a continuity between philosophy and the sociology of science. Used in this way they serve only as a form of self-legitimizing. Recalling illustrious ancestors is the simplest means of reinforcing the belief in the existence of a genuine tradition of research!

At this point we can formulate some concluding remarks. First, the genealogy of relativism is to be found more easily in hidden antiscientific readings than in the texts that we have examined in this chapter.[13] Second, their frequent recourse to "authority" is perhaps a mark of the weakness, not to say the failure, of their programs.[14] Third, this suggests that the SSK may yet rediscover a path to the rational analysis of scientific knowledge.[15]

Notes

1. It is not the purpose of this chapter to remind the reader of these difficulties; see Chapter 1, note 9.
2. In his time, Robert K. Merton (1953) makes a similar criticism of functionalism: *"Too often we have used, either a single word to represent different concepts, or various words to translate the same concept.* Clarity of analysis and accuracy of expression have suffered from this ill-considered use of words" (68, italics mine). Bunge (1991) expresses even more radical doubts about the new sociology of science: "An ideological program is a confession of faith and a plan to reinforce and propagate the faith. A scientific program is a research project that starts with problems, not principles other than the general philosophical principles underlying all scientific research—for example, that the external world is real, lawful, and knowable" (537).
3. As Anastasios Brenner noted (1990: 36–43), Duhem's works are strikingly consistent on this point. The conception appears in two articles published in 1894, "Quelques réflexions au sujet de la physique expérimentale" and "Les théories de l'optique," where Duhem writes: "An experiment can never condemn an isolated hypothesis but only a whole theoretical group" (1894a, 187) and "The physicist can never subject an isolated hypothesis to experimental control, but only a whole set of hypotheses" (1984b, 112).
4. In several places Duhem (1991) provides a definition of the word "hypothesis": "We connect the different sorts of magnitudes, thus introduced, by means of a small number of propositions which will serve as the principles in our deductions. These principles may be called 'hypotheses' in the

etymological sense of the word for they are truly the grounds on which the theory will be built" (20). Also: "Even if the word 'hypothesis' had in ordinary discourse acquired this sense of dubious assumption, philosophers and astronomers preseved its etymological meaning, namely, that of a basic proposition on which a theory rests" (1969, 101).

5. On this point, it is perhaps unjustified to perceive a radical break between *La théorie physique* and *Sôzein ta phainomena*, as is suggested by Petroni (1994, 114n). The developments expounded in 1908 are in perfect agreement with the thesis of the representation of the real that appears in his first book. Compare for instance, "Astronomy cannot grasp the essence of heavenly things. It merely gives us an image of them. And even this image is far from exact: it merely comes close. Astronomy rests with 'the nearly so'" (Duhem 1969, 21) and the passage in which he (1991) compares explanation and representation (34). The only discrepancy concerns the formula "save the phenomena." Note that the Greek *sôzein* can mean equally: "save, *keep in memory*, observe," and that an equivalent word from Plato is: *diasôthènai*, which also means "*keep faithfully*." Therefore, the title *Sôzein ta phainomena* could have been understood in less emphatic terms. Duhem simply uses the expression to convey that astronomers have often preferred to discuss phenomena rather than theories.

6. Duhem (1991) writes: "Now it may be good sense that permits us to decide between two physicists. It may be that we do not approve of the haste with which the second one upsets the principles of a vast and harmoniously constructed theory, whereas a modification of detail, a slight correction, would have sufficed to make these theories accord with the facts. On the other hand, it may be that we find it childish and unreasonable for the first physicist to maintain obstinately at any cost, at the price of continual repairs and many tangled-up stays, the worm-eaten columns of a building tottering in every part, when by razing these columns it would be possible to construct a simple, elegant, and solid system" (217).

7.   Let us add that, in the context of the verificationism admitted by Quine, the meaning of a theory lies only in its procedure of verification.

8. Quine considers such holism unjustified. He (1960) writes: "This point has been lost sight of, I think, by some who have objected to an excessive holism espoused in occasional brief passages of mine" (11).

9. As for the first (observational statements), they are evidently revisable, but only within certain limits. Here Quine (1980) is a proponent of realism: "We cannot significantly question the reality of the external world, or deny that there is evidence of external objects in the testimony of our senses" (220). This invalidates principle R1 of relativism.

10. More accurate propositions have been put forward by Wright (1992) and Engel (1998). In the context of this minimal conception, the truth would be, as Engel (1998) explains, an *overassertibility* norm, "because it records that our assertions are justified in the way that is most stable, absolute, and shielded against revision. Even if we are never certain to have reached this, perhaps mythical, ideal limit, it is what we aim for" (72).

11. We should deplore carelessness with which Klossowski's edition was prepared. There are errors in the numeration, of which some of the most serious are the following: 2.051 (= 2.0251), 2.063 (= 2.062), 3.31 (= 3.031),

521 (= 5.521), etc. They make it very difficult to follow Wittgenstein's chain of reasoning.

12. When asking questions such as: On what basis does the respect for rules stand? Does the breaking of rules involve a penalty? Are rules learned? etc., we rediscover large parts of the Mertonian sociology of science.

13. Beginning with the works of Nietzsche, Spengler, and Heidegger. A study of the careers of the most influential scholars in the humanities reveals an entire generation that was marked by Nietzsche and his commentators. Some would later deny this influence on their work, while others consolidated their skeptical and nihilist leanings. In the field of SSK, Latour (1997) openly admits the influence of Nietzsche on his conception of science (125), and Fourez (1996) refers to the German philosopher in various passages (35–37, 363–366). Regarding Nietzsche's reception in France, see Staszak (1994). We also find in Bouveresse (1973, 1984) some elements that shed light on the contemporary spread of relativism.

14. Siegel's critique of relativism (1987) is discussed in Chapter 1. He put forward two main arguments, the first being the UVNR argument (relativism undermines the very notion of rightness), and the second the NSBF argument (necessarily some beliefs are false).

15. In addition to studying styles of thinking, SSK addresses the question of statements that are wrong (logical lacunae, observational errors) and statements that are untestable (axioms, postulates, *ad hoc* statements). Untestable statements form only a marginal part of a scientific theory, because their truth-values cannot be determined. Strictly speaking, they are "a-scientific statements." This is why SSK seeks to understand whether social interests or beliefs may lead to the adoption or rejection of a scientific theory.

Conclusion

Toward an Epistemological Incrementalism

The scientific controversies that have been examined here allow us to glimpse results that are quite different from those reached by sociological relativism and constructivism. In a word, nothing that has been revealed by these controversies—whether drawn from different periods of institutionalized science (Chapters 2 and 3) or even much earlier from medieval Islam and Latin Europe (Chapters 4 and 5)—contravenes the classical principles of science, beginning with ontological realism (the world exists), epistemological realism (the world is knowable), and methodological rationalism (reason is the best way to know the world).

The discrepancy between the results of our empirical enquiries and those of SSK studies stem primarily from questions of *methodology*. As we have seen in these case studies, the analysis of scientific knowledge will continue to encounter difficulties:

1) so long as research on the *asymmetries* between the actors in a controversy is not conducted in a systemic manner. The presumption that there is a relationship between the production of scientific results and their social milieu sometimes stems from the fallacious extrapolation of personal correlations that taken together shed little light on how science actually functions.
2) as long as the concept of *determination* persists and is not replaced by a spectrum of relationships within which one can discriminate between conditioning, interdependence, functional correspondence, congruence, etc. The term "determination" comprises and confines these relationships within a rigid framework of efficient causality, which presents here the same inexactitudes as the cruder versions of social determinism.
3) as long as SSK continues to embrace a simplistic view of *rationality*, which can lead to the confounding of *rational statements* with the *norms of rationality* that regulate the production of such statements, and to the relegation of any idea that has not been analyzed in this way to the domain of beliefs.
4) so long as the notion of *convention* is applied to undifferentiated content (facts, methods and theories) rather than being linked to procedures or to nontestable statements (axioms, assumptions, *ad hoc* hypotheses, etc.).

5) as long as one attempts to impose the notion of the *social negotiation of truth* on the production of scientific knowledge, a notion that springs from a political model of action and ignores the highly differentiated nature of scientific activity.

6) as long as one continues to view the relationship between scientific production and the reference texts of the philosophers of science from the perspective of *self-legitimation* rather than choosing to follow the authors' arguments. On this point, the relativist positions stem from a radicalization of thought.

The conclusions that we can therefore draw with regard to the problem of scientific controversies are of two orders—the sociological and the epistemological. The first conclusion sets out a model along the lines of structural methodological individualism, while the second aims to show that scientific controversies could provide keys to the epistemological debate with regard to falsificationism and verificationism.

1. Sociological Conclusion

Let us return to an aspect of scientific controversies that was underlined in the introduction, and which consists in regarding all controversies as a separation within the scientific community in the interpretation of a given phenomenon. From this point of view, all scientific controversies depend on a *separate* scheme (Raynaud 2006, 73–78),[1] which in the case of bipolar debates is a scheme with two actants.[2] From the perspective of classical sociology, it can be assumed that the participation at a time t of an agent in a given action requires a configuration that is favorable to his interests and values. This provides, it may be said, the motive for the appearance of "stable configurations," which leads us to examine the specific structure of the interests and values that could intervene in a scientific controversy.

1. Interests and Values

Let us begin by addressing two fundamental objections that have been raised against the model of controversies under consideration here. These objections focus on the values inherent in the model and the way in which the model can be adapted to the specificity of different scientific controversies.

1. Why should values be introduced into a model of controversies? Studies of scientific judgments based on the utilitarian tradition only consider the interests of the individual agents. The model of rational choice as applied to scientific activity (Radnitzky 1987) assumes that scientists will prefer option A to option B if and only if $ue_A > ue_B$, in which ue is the expected utility. When applied to the problem of controversies, this decision model could explain why a scientist may accept or refuse to pursue a controversy with an adversary.

However, the model suffers from various lacunae. It is known, for example, that decision models face epistemological hurdles such as Hempel's problem of induction, the problem of known data, etc. (Boyer 2000, 172–175). A first step was made toward overcoming these difficulties by Giere (1988), who applied the model of bounded rationality (Simon 1983) rather than expected utility to scientific judgments. He assumed that, because they will only have access to a limited quantity of information in a limited space of time, the agents to a controversy are not in a position to maximize the function of utility; the most they can hope to achieve is a "satisfying" suboptimal solution. Some sociological studies have gone even further. Raymond Boudon (1995, 349–350) demonstrated through numerous examples the limits of the utilitarian theses. For example, theft is undeniably useful if one considers its results, and one may therefore ask why it is not socially encouraged. The response is obviously that theft violates the principle of the correspondence between contribution and retribution and therefore leads to *moral disapproval*. This is a typical case in which the judgment of the agents is based on irreducible axiological motives within a utilitarian framework.

Similar examples can be cited in the realm of scientific activity. While it is evident that a physician has an economic interest in there being a large number of sick people, it would be absurd to suppose that he might seek to propagate viruses within a population in order to satisfy his cupidity. The *moral* duty of a doctor is to care for his patients and the Hippocratic Oath exists in order to remind him of the axiological values to which his profession is bound. In the same way a savant has a manifest interest in publishing a large number of papers, for this is how he can gain increasing credibility and recognition. However, the scientist is held to the moral requirement that he only publish findings which are consistent, substantiated by experimental tests, and produced in accordance with deontological norms. Great scientists are credited not only for their discoveries, but also for their moral qualities in order to remind us of the axiological foundations of their work. It is only in *limiting cases* that some scientists decouple their personal interest from the requirements of moral judgment. Well known examples include John Long (1979), Paul Chu (1987), Viswa Jit Gupta (1991), Friedhelm Herrmann (2000), Scott Henninger (2007), and Marc

Hauser (2012). Therefore, both personal interest and universal values are mobilized in scientific controversies.

2. In what way do scientific controversies differ from other forms of conflict? A model based on interests and values has broad applications. It can be used to represent a private dispute, an episode of economic competition, a political disagreement, or even a war. But does this challenge our previous assertion that scientific activity is highly differentiated and that scientific knowledge is universal, robust knowledge? The model of interests and values applied to *separate* schemes can embrace all manner of conflicts. However, a given combination of interests and values will always be specific to the sphere of activity under consideration, and the importance of this set of interests and values will vary from one sphere to another. While one may accept that economic interests predominate in a case of commercial competition, it would be unreasonable to assume that they will play a role in a theological debate. In the same way the ideological values that lie at the core of a political debate will play only a secondary role in a private dispute. The combination of interests and values is *dependent* on the context of the action.

The problem then becomes to determine the nature of these interests and values in the characterization of a scientific controversy. It may be suggested that such disputes call into play specific *cognitive interests and values* that do not characterize any of the other conflicts listed above. It is difficult to find strict equivalents between a conflict unfolding in the private, political, religious, or economic sphere and the situation of a researcher expressing doubts concerning a methodology (cognitive interest) or taking up the defense of a theory that offers an elegant synthesis of a set of results (cognitive value). Noncognitive values and interests clearly exist in the sciences, but these are inhibited[3] by the interests and values associated with knowledge, except perhaps in those cases where scientists renounce their autonomy in favor of a party or lobby. Conspicuous examples of this may be cited: the raciology of Fisher and Schickedanz under Hitler, the anti-Mendelian genetics of Lyssenko under Stalin, and so on.

The priority of cognitive values and cognitive interests is sufficient to differentiate scientific controversies from other forms of conflict.

2. Cognitive Interests and Values

Identifying the different classes of motives that exist for those taking part in a controversy is of great utility as it allows us to demonstrate the multifactorial nature of scientific debate and to resurrect conclusions that may have been abandoned but remain valid (Collingridge and Reeve 1986; Collingridge 1989).

Let us review first of all the conclusions of Collingridge and Reeve (1986) regarding the formation of a consensus. They identified three necessary conditions for a scientific consensus to exist: (c1) the research must be autonomous vis-à-vis the interest groups involved; (c2) the problem must pertain to a single discipline; and (c3) the level of criticism must be relatively low. Since controversy and consensus exist by definition in a state of opposition, it is possible to translate the conditions for scientific consensus into the conditions for a dissensus, (i.e., for a scientific controversy). It suffices to consider that a controversy exists only if one of the three conditions $\neg c1$ or $\neg c2$ or $\neg c3$ is met.

Condition $\neg c1$ states that a controversy can be conducted by lobbies or interest groups. It is necessary to fill the gap by pointing out that *opposite* interests are involved here, because a controversy obviously cannot arise when the interests of the parties are congruent.

Condition $\neg c2$ signifies that controversies arise more easily when different disciplines come into play. However, numerous studies have shown that researchers in different disciplines have different cognitive interests.

Condition $\neg c3$ can be interpreted in one of two ways. On a more superficial level, one could say that a controversy is the product of one's critical capacity. On a deeper level, since one's critical capacity is a personal predisposition, a controversy has a greater chance of developing when the scientists share an acute critical sense and therefore a predisposition—*nolens volens*—to engage in a controversy.

The multifactorial nature of scientific controversies becomes clear if one is willing to entertain two modes of thought.

First, let us imagine a scientific community in which the interests and values of the actors are strongly antagonistic *even if the knowledge in this discipline is consensual* (if, for example, no rival theories exist). It is clear that a situation involving a mere antagonism between values and interests is not sufficient to create a controversy, as it would have no scientific foundation. An essential condition is lacking: the existence of rival hypotheses. The classic SSK model, which only recognizes the extrascientific values and interests of the actors, cannot explain the absence of controversy in such cases.

Now let us suppose the inverse: that the knowledge produced by a discipline is highly contradictory, *but the interests and values of the members converge.* Imagine a situation in which the scientific community has every reason to avoid controversy (perhaps because it includes a large number of unproductive researchers) and at the same time it shares a high congruence of interests (the members have all undergone a similar formation, and are working on the same problem using the same methodologies). This combination of interests and values makes it likely than any controversy will be quite weak, even if the researchers are defending opposing hypotheses. The classical epistemological model based on contested knowledge cannot be applied to this type of situation, which hardly deserves to be called a controversy. Any debate will be of short duration with a low degree of conflict, and the researchers will probably end up collaborating or exchanging results. The adhesion of the researchers to opposing hypotheses does not necessarily give rise to controversy.

Factors that are sociological in nature are therefore only *indirectly* linked to the hypotheses that are under debate in a scientific controversy. The sociological dimension depends more on the *form* that the debate may assume, that is to say, the more or less conscious choice that the actors may make to engage in a *confrontation* rather than to accept the *coexistence* of rival hypotheses or to *support* one of them. Here only the degree of antagonism (i.e., the propensity of a community to foster division) is social in nature and must be explained sociologically. The existence within a discipline of rival hypotheses is no more than *a precondition for the possibility of a controversy.* Considered in isolation, an analysis of the *content* of rival hypotheses cannot explain why it may engender courteous discussion in one case, bitter dispute in another, and in some cases no clash at all between opposing researchers. Sociology will never be able to provide a response to these questions unless it deepens its understanding of the methods of epistemology.

Relativism and constructivism often proclaim that objectivity is an illusion and that there are no differences between scientific knowledge and political or religious belief. In the strongest versions of relativism and constructivism, any influence of the objective world on the production of scientific knowledge is denied. In the weaker versions, the existence of objectivity is not denied, but it is defined as "intersubjectivity." Some sociologists who are critical of the relativist position nevertheless support the idea of intersubjectivity based on Popper, who laid the groundwork for contemporary relativism (Popper 1979, 2:149–151). They point out, for example, that "to refer to the objectivity of science is just another way, from the sociological perspective, of referring to the intersubjective agreement between the members of a scientific community" (Gingras 2013, 87). But is this really so?

Two objections can be advanced here. Impartial examination will show that the facts occupy the four cells of a contingency table: (1) knowledge can be true and based on consensus, as in the law of refraction; (2) it can be true but nonconsensual, as with the theory of relativity in the 1930s; (3) it may be untrue but consensual, as with Mitchourinism in the USSR; or (4) it may be both untrue and nonconsensual, as for example in the thesis of water memory claimed by Jacques Benveniste. The random occupation of the four cells by these facts shows that *scientific truth bears no relationship to the intersubjective accord between researchers*.

This objection would be sufficient in itself, but it can also be extended if one explores the assertion that objectivity is equivalent to an intersubjective accord. There are a number of authors who have developed an "intersubjective accord" with regard to the contrary thesis, which was expressed as early as the seventeenth century by the precursors to the Enlightenment known as the *"Rationaux."* In *Regulae ad directionum ingenii* (1628) Descartes writes: "It is pointless for us to count the votes, in order to determine the opinion that claims the greatest number of adherents" (1826, 210). Gilbert Naudé came to the same conclusion: "One must not in any way hold our liberty captive to the number of votes and induce ourselves to approve like a band of paltry judges (*juges pédanées*[4]) everything that it pleases them to declare to us" (Naudé 1653, 5). Pierre Bayle is even more clear on this point: "A sentiment cannot be accepted as probable for the multitude of those who follow it, except insomuch as it appears to be true to many, independently of all prejudices and by the sole force of judicious examination" (Bayle 1683, 134). It is now generally agreed that the Enlightenment philosophers held the view, as stated by Helvetius: "Experience must convince us that a multitude of adversaries can prove nothing against the truth of a given proposition" (Helvetius 1775, 61). Piaget contested the conclusions of Durkheim with the same acuity, declaring: "[. . .] insofar as the truth rests on an accord between (like) spirits, it has been [mistakenly] concluded that any accord between (like) spirits engenders a truth" (1965, 82). Piaget had no difficulty demonstrating the paralogism. One cannot infer an intrinsic property of knowledge based on its behavior in the social space.

As a consequence, the relativist and constructivist theses, even in their weak form, that scientific knowledge cannot attain objectivity are quite improbable; nothing allows us to leap in one bound from a belief held by consensus toward objective knowledge. Returning to the methodological level, this thesis should at least be supported by a program of comparative analysis which shows that the procedures for the construction of scientific statements and social beliefs are identical. Since they are in fact fundamentally different, one can understand why such an analysis has never been undertaken.

2. Epistemological Conclusions

Scientific controversies offer a second point of interest for social scientists, one that lies at the intersection between epistemology and the sociology of the sciences. In fact, it appears that the scientific controversy can provide fresh arguments for the debate that began in the early twentieth century between the logical positivist thesis of *verificationism* and the critical rationalist theory of *falsificationism*. The latter challenged the very notion of empirical science and the scientific method and appears to have emerged victorious, if one is to judge by standard contemporary epistemology which globally reflects the work of Karl Popper.[5]

1. The Falsificationist Thesis

The moving force behind logical positivism was the Vienna Circle, a group of philosophers who, drawing upon the work of Wittgenstein, promoted *verification* as the criterion of demarcation between science and nonscience; every statement that can be verified by proofs is scientific. The logical positivists advanced a semantic theory according to which "the meaning of a statement is its method of verification" (in this formula one can recognize the influence of Wittgenstein on Schlick and Waissmann). However, in the face of criticism by epistemologists, Carnap (1936) weakened the theory by changing verification to *confirmation*.

As early as 1934, Popper independently contested the positivist criterion of verification, based on the conviction that the proofs marshaled in defense of a theory can never be definitive (Popper 1973, 29). Any theory "verified" at a time t_0 remains subject to revision at a later time t_1. The conundrum of induction was thus revived, allowing Popper to actively criticize the inductivist theses.[6] In his famous thought experiment, the fact that one has seen a thousand white swans does not allow one to conclude that "All swans are white," because a solitary observation to the contrary is sufficient to invalidate the proposition based on a thousand previous observations (Popper 1973, 23). One thousand white swans does not verify the statement "All swans are white," whereas a single nonwhite swan is sufficient to falsify "All swans are white." Such is the asymmetry between verification and falsification.[7] One can demonstrate the *falsity* of a theory but not its *truth*. One of the consequences deduced by Popper from this asymmetry was that the sole acceptable criterion of demarcation is the falsifiability of a theory (Popper 1973, 36–39 passim).

Falsificationism swept away such positive formulas as: "Theory T has been verified" or "established" or "proven." These were forced to give way to negative statements such as "The theory has not been falsified" or "invalidated," etc. Other, more subtle and profound consequences of falsificationism were drawn by Popper and his disciples, but these lie beyond the scope of these lines.[8]

2. Falsificationism versus Verificationism?

On the other hand, what we must examine here is the triumph of falsificationism, which has massively discredited the thesis of verificationism and prevents those who wish to challenge the limits of Popper's negative epistemology from reverting to the arguments of the Circle of Vienna.[9]

Where can we begin? One first demonstrable point is that there are cases in which falsificationism and verificationism are not inherent enemies. For reasons completely extraneous to the questions being explored here, and based on an analysis of situations that lend themselves to the engendering of false ideas,[10] Boudon has drawn attention to the role of the *logical framework* within which a question is formulated (1990, 184–188). He noted that the persuasiveness of Popper's thesis rests on the implicit hypothesis that scientific theories exist in an open environment and writes: "Paradoxically, it is because he adopts a very *liberal* hypothesis with regard to the logical environment of the questions posed by the scientist that Popper himself introduces a logical framework that is overly restrictive and inadequate in relation to reality [. . .] In fact scientific activity is *open* enough to pose *closed* questions as well" (1990, 186). Science does not usually concern itself with closed questions, but these offer a more interesting case of reasoning than would appear *prima facie*.

As we all know, the great scientific controversies have generally been bipolar in form, with two actors or parties defending rival hypotheses. It is also known that in a bipolar controversy there is a tendency to convert these *rival hypotheses* (*p* or *q*) at a very early stage into *opposite hypotheses* (*p* or ¬*p*), which flies in the face of Duhem and Popper's opinion that genuinely opposite hypotheses cannot exist. Scientific controversies nevertheless provide a wide array of bipolar hypotheses. For example, before the first observations of the eclipse of the moons of Jupiter by Rømer in 1676, some believed that light propagates instantaneously and others that light *does not* propagate instantaneously (Cohen 1939). Analogously, in their well-known *querelle* Saint-Hilaire argued that there was a single basic plan of organization for all animal forms, whereas Cuvier stated that such an organizational plan *did not* exist. In the debate over spontaneous generation, supporters of the thesis that microorganisms are born of parent microorganisms opposed supporters of the thesis that microorganisms *are not* born of parent microorganisms. In one of the fundamental debates in the history of chemistry, Prout sustained that the atomic weights of all elements were multiples of a unit weight (the atomic weight of hydrogen) against Stas, who argued that they *were not* multiples of a single unit weight (see Chapter 5). In the "Battle over the Electron," Ehrenhaft was convinced that he had discovered smaller electric charges than the charge of the electron, but Millikan succeeded in demonstrating that such charges *did not* exist (Holton 1982). In the "Great Debate" of the 1920s between astronomers regarding the size of the universe, Curtis advanced the thesis that the galaxies (then referred to as "nebulae") were "island universes" located outside the Milky Way, while Shapley believed that this *was not* the case (Hoskin 1976). All of these controversies, and a host of others that can be drawn from the history of the sciences, utilize the same logical framework p w $\neg p$, in which the rival hypotheses are reduced to opposite hypotheses.

Scholars often skip over the fact that the philosophers of the Enlightenment arrived at a similar diagnosis on comparable subjects. The marquis de Condorcet, whose work has been thoroughly evaluated by Granger (1956) and Rashed (1974), left illuminating notes on the subject of controversies, the fruit of his lifelong interest in the notion of public opinion.[11] Condorcet observed and described the process of reducing different hypotheses to opposite hypotheses, writing in his *Essay on the Constitution*:

> In many assemblies, the topic of deliberation is only sketched out in general terms; it can generate as many opinions as there are heads, or at least people who can or wish to become opinion leaders. Gradually, by dint of discussion, the various opinions will be reduced to a smaller number; and in the long run, with everyone reuniting voluntarily or being obliged to unite in accordance with the established rule, in support of one of the two dominant opinions, one of these opinions will end by having the majority (*pluralité*) behind it [. . .]. But after having

examined the different forms of deliberations that can be employed, one will find that there exists only one way to obtain, on a subject that has been submitted to an assembly for their decision, the true will of the majority (*pluralité*) or, in what comes to the same thing, the decision whose truth is the most probable. This method consists in reducing [these opinions] to simple propositions, on which one can only vote yes or no (1788, 116).

It would be no betrayal of the ideas of Condorcet to assume that what holds true for assemblies also holds true, at least in part, for the scientific debate, because in developing his analysis of the "reasons to believe" (*motif de croire*), Condorcet's scope was not just to reform the institutions. He was also elaborating a theory of the human spirit that would permit him to establish a theory of knowledge. In his analysis, the honest man and the savant both reason in the same manner when they embrace an idea. This is stated explicitly in *Éléments du calcul des probabilités*: "The reason we believe in facts regarding which we only know the probability,[12] is the same reason that leads us to believe the truths according to which we act, we regulate our conduct, and *we reason in the sciences*" (1974, 133, italics mine). Despite the differences that we have underlined, Condorcet's examples share some of the traits of the scientific debate. In each case the protagonists are searching for a stable solution and the establishment of opposite hypotheses represents the most efficient method to arrive at such a solution.

And yet it is precisely in such situations that the falsificationist and verificationist theses coincide. If the logical framework of the question posed takes the form $p \text{ w } \neg p$ (w standing for "where exclusive"), one can then apply the following deductive rules:

$$
\begin{array}{ll}
\dfrac{\begin{array}{l} p \text{ w } \neg p \\ \neg(\neg p) \end{array}}{p} & \qquad \dfrac{\begin{array}{l} p \text{ w } \neg p \\ \neg(p) \end{array}}{\neg p}
\end{array}
$$

An experimental test will then provide the response with certainty—either p or $\neg p$—and one of the two theses will have been "verified" because the other has been "refuted." This idea is foreign to the philosophy of Popper (1991, 148), which only admits the existence of a choice between two hypotheses on the basis of a "rational preference." It is equally foreign to the critique of the crucial experiment made by the post-Popperians under the influence of Duhem. Condorcet's notion of "*motif de croire*" remains pertinent only so long as the problem fits into a logical framework of the type p w $\neg p$. One can nevertheless attain the certitude each time that:

1) the hypotheses are all mutually exclusive;
2) the hypotheses are limited in number;
3) all of the hypotheses are refuted with the exception of one.

The possibility that one may reduce a complex logical framework $(p$ w q w r w ... w $z)$ to $(p$ w $q)$ by means of an ordered series of controversies offers a clear guarantee of simplicity in the search for the truth. The difficulty resides in the implicit terms of reasoning. As Boudon perspicaciously noted (1990), inconsistency in reasoning can arise from an error in the comprehension of the logical framework.

Two obvious examples can be cited.

Case 1. The faulty logical framework does not ruin the conclusion. The debate over whether the Earth is a sphere has historically been presented in the form of two possibilities: Is the Earth round or is it flat? In truth the question should take the form p w q w r: Is the surface of the Earth convex or concave or is it a flat plane? The hypothesis of concavity being immediately refuted by the existence of the horizon, one could legitimately limit the question to p w q. Eratosthenes of Cyrene (c. 276 BC–c. 195/194 BC) found the solution by studying the shadows cast by two vertical columns, one erected in Syena and one in Alexandria. At noon on the summer solstice, the column in Syena did not cast a shadow (the sun being at its zenithal culmination) whereas the column in Alexandria did (the sun being in simple culmination). Whence the valid conclusion $\neg q$ therefore p.

Case 2. The faulty logical framework does ruin the conclusion. In *Maqāla fī daw' al-qamar (On the Light of the Moon)* (Masoumi, 2006, 3–92), Ibn al-Haytham proposed a theory to explain the source of the light of the Moon. After reviewing the interpretations of the classical authorities, he presented a series of alternatives: either (p) the Moon is self-luminous or (q) it receives its light from the Sun. In the latter case, one must then ask how the Moon receives this light: (q_a) by reflection in the same way as mirrors, (q_b) by refraction as in transparent bodies, or (q_c) by some other means? If by some other means, then: "The body of the Moon must, when it is lit by the Sun, become self-luminous *(mudi'a min dhātih)* [. . .]. The light then emanates from the entire body of the Moon, as it emanates from luminous bodies in themselves, and not as it does from bodies that transmit light by reflection *(ini'kās)* or by refraction *(al-nufūdh)*" (Masoumi 2006, 27).

Ibn al-Haytham then embarks on a series of refutations: p (light emanates from the Moon) is not possible, otherwise the Moon would shine even during lunar eclipses; q_a (the light of the Moon is reflected) is refuted because if this were true one should see images on its surface as one would in a convex mirror; and q_b (the light of the Moon is refracted) is clearly untrue because otherwise the Moon could not hide other celestial bodies, as it does during solar eclipses. In this way our astronomer arrived at thesis q_c, explicitly describing his reasoning as $(q_a \text{ w } q_b \text{ w } q_c) \neg q_a \neg q_b$ then $\neg q_c$: "The light emanating from the Moon is not due to reflection. We have demonstrated that it is not due to refraction either. Therefore, among the modalities according to which light may emanate from bodies, there only remains one: the Moon is luminous in itself because it is lit by the Sun" (Masoumi 2006, 91). Ibn al-Haytham nonetheless had to ask himself whether he was utilizing a faulty logical framework ($q_a \text{ w } q_b \text{ w } q_c$). Indeed, it would not be until the eighteenth century that savants (Lambert *Photometria*, 1760) defined another modality for the transmission of light—diffuse reflection—in which no image appears on the surface of the Moon. Ibn al-Haytham's logical framework could then be correctly rewritten as $(q_a{}^* \text{ w } q_a{}^{**} \text{ w } q_b \text{ w } q_c)$, with $q_a{}^*$ standing for specular reflection and $q_a{}^{**}$ for diffuse reflection.

3. Toward Incrementalism

If we accord some degree of validity to the developments described above, it is because bipolar controversies are a *regular* occurrence in the history of the sciences. Nevertheless, Popper is correct when he observes that on a long timescale the universe of solutions is completely *open*. The preceding observations cannot justify a return to the classic form of verificationism; they simply indicate that verification may be pertinent in the treatment of certain problems.

If verificationism and falsificationism are not inherently in opposition in the short term (since the two theses are equivalent within the logical framework $p \text{ w } \neg p$), it remains to be determined whether they are incompatible over the long term. Quite surprisingly, the response appears to be negative. If one considers the aftermath of controversies on a given scientific question, one will note that the repeated clashes between opposing hypotheses often result in the elimination of a false hypothesis for a hypothesis that is not only *more acceptable*, but also *perfectible*. Popper, who granted the validity of this evolutionary model of the truth in general terms, drew the conclusion that any certainty acquired at time t_0 is erroneous and therefore simply an illusion. Here he was adopting a maximalist position. With regard to the objection that, "The old theory, even when it is superseded, often retains its validity as a kind of limiting case of the new theory," he responded laconically, "There is something to be said for the above argument, but it does not affect my thesis" (Popper 1992, 249–250). There is, however, another less logicist and more direct way of responding that permits one to draw up an alternative model of scientific progress based on controversies.

However, we must then answer the question: "In what way are certitude and the perfectible solution compatible?" Let us grant that the logical framework within which a question is posed never coincides exactly with the structure of the real, but depends on the data available at the time t_0 when the question is asked. It is entirely possible that at a later date t_1, some observational or some theoretical data (case 1 and case 2, respectively) will justify a rearrangement of the logical framework within which the problem has been posed.

When geographers at the end of the seventeenth century rediscovered Eratosthenes's proof, they acquired the certainty that the earth was round, but they then had to define the concept of "roundness": Was the earth actually a sphere or a flattened ellipsoid? Based on an observation of Jean Richer regarding the variation in the length of a pendulum used to measure the second as a unit of time, Newton and Huyghens almost simultaneously came to the same conclusion that the earth was a flattened sphere. This hypothesis was confirmed in 1737 by the geodesic measurements made by Maupertuis during a scientific expedition to Lapland (case 1: theoretical reformulation).

In the same way, based on the early notes of Galileo and the observations of the eclipse of Jupiter's satellites by Rømer, savants arrived at the certitude that the speed of light was not infinite, although doubts regarding the values calculated by Rømer remained. In *Adversaria* Rømer declared that light traveled 1,091 earth diameters per minute, whereas in a work published on December 7, 1676, in *Journal des Sçavans* he stated that the diameter of the Earth was 3,000 lieues without specifying the unit measure he had adopted—whether Parisian lieues or common lieues. In *Comptes-Rendus de l'Académie des Sciences*, Fontenelle indicated that Rømer had measured a speed of light of 48,203 common lieues of France per second, or approximately 214,000 km/s. Doubts were later expressed regarding this value. Francois Arago invented a device (subsequently perfected by Fizeau and Foucault) which raised the hypothesis of a high value for the speed of light that was confirmed by Fizeau in an experiment using a toothed wheel (315,300 km/s), but which was weakened by Foucault's demonstration using a rotating mirror (298,000 km/s) (case 2: empirical reformulation).

These two cases suggest the existence of a somewhat awkward form of context-dependent certainty. Does this not constitute a *contradictio in adjecto*? There is at least one reason that justifies our referring to a certainty rather than to the false illusion of certainty (in line with the Popperians[13]) or local belief (as per the relativists): It is that certain hypotheses (such as the flatness of the earth or the infinity of the speed of light) can be eliminated once and for all. Nothing prevents us from concluding that a *definitive* response to the question p w $\neg p$ may not also be *perfectible*. *Stricto sensu* the speed of light calculated by Rømer was not "falsified" by Fizeau and Foucault; it was *rectified*. The roundness of the earth was not "falsified" by the discovery of the flattened ellipsoid: it was merely *rectified*. Beginning in the 1960s satellites circling the earth have allowed scientists to detect and measure a slight depression at the South Pole and a bulge of the same magnitude (about 17 km) at the North Pole. The model of a flattened ellipsoid has therefore itself been *rectified* by the model of the pear-shaped ellipsoid.

On more complex problems, progress in the sciences may gather momentum from a succession of controversies, each one of which attempts to reduce the space occupied by the range of possibilities that existed at an earlier time. A controversy p w $\neg p$ making its appearance at time t_0 could provoke a new controversy q w $\neg q$ at time t_1, which could itself give rise to a controversy r w $\neg r$ at time t_2, and so on. At each stage in this sequence the response may be different, but the process will unfold within an increasingly restricted space of possibilities. Contrary to Popper's negative epistemology, this model of knowledge *does not grant any space for relativism, skepticism or nihilism* such as developed in the school of SSK. Over the long term the knowledge of a problem would appear to advance by means of the exploration of a logical tree of alternatives. Moreover, no scientist can claim predominance over others in determining the direction of this exploration. Scientific hypotheses and the methods of exploration of a problem cannot be planned. This could form the basis for an *epistemological incrementalism*.

In its most general version, incrementalism refers to a form of collective action in which the transformation of organizations or their products depends on a series of small, unplanned changes (Brayrooke and Lindblom 1963; Quinn 1984).[14] While each of these changes may be imperceptible, their gradual accumulation can lead to a radical change.

Collingridge (1989)[15] applied these ideas to the study of technologigal discoveries, showing by means of two case studies—the development of aseptic surgery by Lister and the perfection of the atomic pile by Pegram, Fermi and Szilárd—how important technological discoveries can be arrived at through an informed process of trial and error (157). It would not be unreasonable to transpose this conception (with due caution) to scientific activity, which would allow one to explain four well-known characteristics of science: (1) the *rational* but *unplanned* nature of its progression; (2) the small, almost imperceptible steps that characterize *normal science*; (3) the *revolution* that comes about when the accumulation of anomalies reaches a level that demands a complete reevaluation and new synthesis of current knowledge; and (4) in the universal history of humanity, the crucial role played by *scientific controversies* in the establishment of truth and the vanquishing of ignorance. This epistemological version of incrementalism applied to the production of knowledge presents two characteristics that deserve to be highlighted before we draw our final conclusions.

First of all, the epistemological incrementalism allows one to envisage the coexistence of verificationism and falsificationism. Typically, in a scientific controversy an accord is reached between two opposing theses because they formulate symmetric results. Within the logical framework p w $\neg p$, refutation amounts to verification. But incrementalism remains equally removed from verificationism and falsificationism, because their meeting point cannot be determined to the advantage of one thesis and the exclusion of the other. All that incrementalism shares with these two theories of knowledge are the fundamental principles that the world exists, the world is knowable, and the world can give rise to true descriptions—explications. Perhaps from this perspective we must await more prudent conclusions than the ones that have been advanced by verificationism and less counterintuitive conclusions than those that have been produced by falsificationism.

Second, by adopting the incrementalist version of epistemology we can effectuate a rapprochement between epistemology and the sociology of the sciences—on the one hand allowing epistemology to describe scientific progress while remaining wary of its penchant for logicism, and on the other hand allowing for the empiricism of SSK while rejecting its delirious episodes of skepticism and nihilism. This rapprochement could prove fruitful, because it allows scientific controversies to occupy the place that is their due in the process of producing truth. Truth and incrementation are not incompatible. The defense of cognitive incrementalism could be viewed as a development of the concept of *critical rectification*, as laid out by the French epistemologist Canguilhem (1977): "Science is a discourse subject to the norm of critical rectification" (21). Instead of viewing controversies as foreign to and little tolerated by science, we have on the contrary good reason to consider controversies as inherent *constituents* of scientific activity, in the sense that they represent the best vector for the critical expression of this activity. This position was brilliantly set out by Marcelo Dascal (2000): "Science manifests itself in its history as a sequence of controversies; these are, therefore, not anomalies but the 'natural state' of science; controversies are the locus where critical activity is exercised, where the meaning of theories is dialogically shaped, where change and innovations arise, and where the rationality or irrationality of the scientific enterprise manifests itself" (165). Whatever label one may assign to scientific controversies—*controversia, contentio, disputatio*, etc.—they have existed since earliest antiquity. Indeed, they have coincided—and not by chance—with the moments of greatest scientific productivity.[16] If scientists regularly clash with one another in debate, it is perhaps because they all know, at least intuitively, that such controversies are the products of a social organization that is extremely effective in constructing robust and stable knowledge of the world. In making this declaration we are not reverting to finalism; rather it allows us to recognize in the social nature of controversies the existence of *proven arguments* to reach the truth.

Notes

1. This brings us to a point where we may envisage a relationship with other works. A controversy will assume the form of a *separate* scheme, just as a social transmission realizes the form of a *send* scheme, the merging of two trade unions realizes the *unite* scheme, and so on. The modeling of an action based on the interests and values that were mobilized in its transmission has been discussed elsewhere (Raynaud 1998, 343–349). This conception of the schemes of action falls into the category of the *formale Soziologie* insofar as it involves: 1) the placing of the actants within a relationship (except in those rare cases of single-actor schemes); 2) an abstract representation of the action, whose realization depends on the existence of interests and values; 3) a form that brings together a wide range of phenomena (a *separate* scheme may manifest in the form of conflict, disputes, agonistic games, controversy, divorce, war, economic competition, etc.).

2. In the present context, I intend the terms *actant, agent,* and *actor* to mean the following: 1) an *actant* is "any participant in a given action"; 2) an *agent* is a human actant; 3) the *actor* is an agent who exhibits particular traits—either individual or categorical—in the action undertaken.

3.   This is why there are not nearly so many scientific controversies as there are political, religious, or other social cleavages. Scientific controversies can only arise in the case of topics on which knowledge and the methods of investigation are well defined.

4.   The *juges pédanées* were itinerate judges without a permanent courtroom, who stood before the public as they handed down their sentences.

5.   By this expression I intend all forms of epistemology that accept falsifiability as a criterion of demarcation. Note that the "research programmes" of Lakatos stem from the idea of the "metaphysical research programme" proposed by Popper in a seminar delivered at the *London School of Economics* in 1949, and here Lakatos freely acknowledges his debt to Popper (1980, 51). Other scholars have captured certain aspects of this notion in their use of the terms "disciplinary matrix," "paradigm" (Kuhn), or "research tradition" (Laudan).

6.   Other points have been subjected to the same critique. Lakatos (1980) reiterated that a factual statement is never pure because it always depends on "observational theories" which, even if implicit, are still active. It is important not to exaggerate this determination of the facts through theories. An observational theory implies that the researcher always interprets his observations by formulating a series of reasonable hypotheses regarding the instruments that he is using to measure the phenomena of interest. To observe a planet through a telescope implies that one has agreed beforehand on the principles of how light is propagated and reflected. These theories are instrumental, and as such play a different role from theories that one is seeking to test through experimentation. Furthermore, observational statements cannot be derived from facts because: "Propositions can only be derived from other propositions, they cannot be derived from facts: one cannot prove statements from experiences—no more than by thumping the table" (1980, 16). The rupture between facts and factual statements must be evaluated fairly. In the passage cited here, Lakatos does not claim that he is negating the existence of any correspondence at all, which would plunge us into a subjective universe where we will be forever imprisoned. The heterogenity between the facts and the factual propositions signifies that, in the act of the falsification of a statement, one can question the link between the fact and the statement by asking oneself whether the hypotheses that were made regarding the relationship between the two need to be revised (critique of the theory of observation).

7.   Boyer observed astutely that Popper was not the inventor of this asymmetry, as it could already be found in the reply of Pascal to the Reverend Father Étienne Noël, dated October 29, 1647.

8.   Within the falsificationist framework, it is clear that the greater the universality and precision of a statement the more easily it can be falsified (Popper 1973, 142). A vague theory is certainly less open to challenge, but this does not constitute proof that it is scientific in nature.

9.   The doubts expressed today with regard to falsificationism appear to reflect not so much a revision as a *weakening* of the thesis following a reorientation of the philosophy of the sciences on questions of regional epistemology. A number of problems have been discovered that do not lend themselves to empirical testability (cosmology, social sciences, etc.) See Boyer (2000, 166).

10. Heinrich Scholz, the historian of logic and heir to Frege, has provided an important element which demonstrates the antiquity of reflections on this theme. In a letter to Gabriel Wagner dating to the end of 1696, Leibniz writes: "Even though M. Arnauld himself expresses the opinion in his *Art de penser* that it would not be easy for anyone to go wrong so far as form is concerned, but solely so far as the material content is concerned, things are in effect quite different. Even Herr Huyghens shared my observation that very commonly the mathematical errors themselves, the so-called paralogismos, arise when the form has deteriorated" (*Phil.* VII, 514–527; Scholz 1961, 51–52).

11. Sociologists typically ignore that Condorcet was the first to draw attention to the "authoritative effects" that became only a focus of attention for contemporary sociology (see Boudon 1986). We can read in *Lettres d'un Bourgeois de New Haven*, Lettre Deuxième: "It would not constitute the sole case where one conducts oneself in accordance with principles that one knows have been demonstrated, even if the demonstration itself is not known. One places one's trust daily in fortune and one's life to a ship's captain or pilot, even when one is well aware that he has no proof of the theoretical truths on which the practical rules that he uses to determine his route are based" (1974, 184).

12. Readers who were interested in this question of epistemology would note that Condorcet provided his own interpretation of the problem of induction. A passage in his work provides a direct answer to the celebrated thought experiment of the "white and black swans" (Popper 1973). Condorcet asked what was the probability of drawing a white card from an urn containing 30,000 cards, of which 10,000 are black and 20,000 are white. The probability of drawing a white card is two out of three. He then asks how one might determine the probability of drawing a white card from an urn whose contents are unknown. His answer was, "I say that it would be by conducting trials [. . .] If, among the three first cards, I draw two white and one black, I could say that there is some slight probability that there are two times more white cards than black. If, among ten cards, four white and six black are drawn the probability increases, and it will increase in proportion to the number of trials or experiments that repeatedly confirm the same proportion of white to black cards. If I conducted three thousand trials, and had two thousand white cards against one thousand black, it would take little [to show] that the probability of drawing a white card was twice that of drawing a black one" (1974, 123–124). There is obviously no guarantee that the drawing of 3,000 cards will reflect the proportion of cards in an urn containing 30,000 cards, but Condorcet demonstrated by means of this thought experiment that the "reason to believe" (*motif de croire*) increases proportionally as one continues to repeat the draw, such that one has reason to convince oneself of the possibility that the distributions in the urn and in one's trials might be identical.

13. Popper later softened his position when he found himself surrounded by a new generation of epistemologists (Kuhn, Feyerabend, etc.). The passage, "It is not impossible to reinforce even the most certain of certitudes [. . .] One may always have a certitude that is even more assured" (1991, 144), takes a step in the direction of the answer we are seeking.

14. I am not challenging the notion of *cognitive incrementalism*, which operates at the crossroads between philosophy and artificial intelligence and refers to continuity between basic perception and abstract cognition. Cognitive incrementalism is defined as "the idea that you do indeed get full-blown, human cognition by gradually adding 'bells and whistles' to basic (embodied, embedded) strategies of relating to the present at hand" (Clark 2000, 135; for discussion, see Downing 2010; Arnau Soler et al. 2013).

15. Collingridge did not envisage any relationship between the incrementalist model and the conduct of controversies, which led him to place incrementalism within the tradition of Popperian epistemology, contrary to what is being proposed here.

16. This fact was clearly understood by the historian of the sciences Ahmed Djebbar: "Polemics of this type emerged above all in the tenth and eleventh centuries, that is to say during the most fertile period of scientific activity in Islamic countries (2001, 88).

Appendices

Appendix 1

The Works published by Pasteur and Pouchet

Reports published in the *Comptes rendus de l'Académie des Sciences* are marked with the symbol°.

Pasteur, Louis (1822–1895)

1847: *Recherches sur la capacité de saturation de l'acide arsénieux*
1848: *Recherches sur le dimorphisme*
1848: *Recherches sur la forme cristalline... et la polarisation rotatoire*
1850: *Recherches sur les propriétés spécifiques... de l'acide racémique*
1851: *Nouvelles recherches sur la forme cristalline... et le phénomène rotatoire*
1853: *Notice sur l'origine de l'acide racémique*°
1853: *Transformation des acides tartriques en acide racémique*°
1854: *Sur le dimorphisme dans les substances actives*°
1855: *Mémoire sur l'alcool amylique*°
1856: *Isomorphisme entre les corps isomères*°
1857: *Études sur les modes d'accroissement des cristaux*
1858: *Mémoire sur la fermentation appelée lactique*
1860: *Mémoire sur la fermentation alcoolique*
1860: *Recherches sur le mode de nutrition des Mucédinées*°
1860: *Note relative au Penicillium glaucum*°
1860: *Expériences relatives aux générations dites spontanées*°
1860: *De l'origine des ferments. Nouvelles expériences... générations spontanées*°
1860: *Suite à une précédente communication relative aux générations spontanées*°
1861: *Leçons de chimie professées en 1860*
1861: *Sur la fermentation acétique*
1861: *Animalcules infusoires vivant sans gaz oxygène libre...*°

1861: *Expériences et vues nouvelles sur la nature des fermentations*°
1861: *Mémoire sur les corpuscules organisés qui existent dans l'atmosphère*
1862: *Quelques faits nouveaux au sujet des levûres alcooliques*
1862: *Études sur les mycodermes. Rôle de ces plantes dans la fermentation...*°
1862: *Suite à une précédente communication sur les mycodermes*°
1862: *Nouveau procédé industriel de fabrication du vinaigre*
1863: *Nouvel exemple de fermentation déterminée par des animalcules infusoires*°
1863: *Note en réponse à des observations critiques... de MM. Pouchet, Joly et Musset*°
1863: *Remarques...*°
1863: *Examen du rôle... de l'oxygène dans la destruction des matières animales...*°
1863: *Recherches sur la putréfaction*°
1863: *Études sur les vins. De l'influence de l'oxygène de l'air...*°
1864: *Des générations spontanées*
1864: *Études sur les vins. Des altérations spontanées ou maladies des vins*°
1864: *Mémoire sur la fermentation acétique*

Pouchet, Félix-Archimède (1800–1872)

1829: *Histoire naturelle et médicale de la famille des Solanées*
1831: *Considérations sur le jardin botanique de Rouen*
1832: *Traité élémentaire de zoologie, ou Histoire naturelle du règne animal*
1832: *Nouvelles considérations scientifiques... sur le jardin botanique de Rouen*
1834: *Flore, ou Statistique botanique de la Seine-Inférieure*
1834: *Introduction à la zoologie antédiluvienne*
1834: *Recherches zoologiques sur la Taupe*
1834: *Expériences sur l'alimentation des animaux par les champignons vénéneux*
1834: *Étude anatomique des globules circulatoires du Zanichella palustris*
1834: *Considérations sur la revivification des noyés*
1835: *Notice zoologique et historique sur les Éléphants*
1836: *Traité élémentaire de botanique appliquée*

1836: *Recherches microscopiques sur la fécule*
1838: *Mémoire sur la structure du vitellus des Limnées*
1839: *Mémoire sur l'organisation du vitellus des Oiseaux°*
1840: *Table pour la revivification de noyés*
1841: *Zoologie classique, ou Histoire naturelle du règne animal*
1842: *Recherches sur la calandre du blé...*
1842: *Théorie positive de la fécondation des Mammifères*
1842: *Recherches sur l'anatomie et la physiologie des Mollusques*
1843: *Mémoire sur l'organisation des zoospermes des Salamandres aquatiques°*
1844: *Recherches sur l'ovulation spontanée et périodique des Mammifères°*
1844: *Sur la progression... du fluide séminal dans les organes génitaux des Mammifères°*
1847: *Anatomie et physiologie de l'appareil digestif du Cousin commun°*
1847: *Théorie positive de l'ovulation spontanée et de la fécondation des Mammifères...*
1847: *Monographie des genres Nérite et Néritine*
1848: *Notice sur les modifications que le sexe imprime au squelette des Grenouilles°*
1848: *Notes sur la mutabilité de la coloration des Rainettes...°*
1848: *Note sur les organes digestifs et circulatoires des animaux infusoires°*
1849: *Note sur le développement et l'organisation des Infusoires...°*
1849: *Note sur l'existence des Infusoires dans le tube digestif des cholériques°*
1849: *Anatomie de la vésicule calcifère des Mollusques°*
1849: *Anatomie microscopique de l'appareil buccal des Nérites°*
1850: *Recherches sur les organes de la circulation... des animaux infusoires*
1853: *Histoire naturelle et agricole du Hanneton et de sa larve*
1855: *Pisciculture. De l'hygiène et de l'alimentation des poissons nouvellement éclos*
1856: *Établissement de pisciculture d'Huningue et du Wolfsbrunnen*
1856: *Lettre sur les bancs d'Anguilles de la Seine*
1856: *De l'hygiène et de l'alimentation des jeunes poissons*
1858: *Histoire naturelle et agricole du Mouton*

1858: *Note sur des proto-organismes animaux et végétaux nés spontanément...°*

1858: *Expériences sur les générations spontanées...°*

1859: *Expériences sur la résistance vitale des animaux pseudo-ressuscitants°*

1859: *Recherches et expériences sur les animaux ressuscitants*

1859: *Étude sur les corpuscules en suspension dans l'atmosphère°*

1859: *Hétérogénie, ou Traité de la génération spontanée* [704 p.]

1859: *Recherches sur les corps introduits par l'air dans les organes respiratoires°*

1859: *Remarques sur les objections relatives aux protoorganismes...°*

1860: *Une expérience à l'air libre sur les générations spontanées°*

1860: *Nouvelles expériences sur les animaux pseudo-ressuscitants*

1860: *Résistance vitale des organismes inférieurs*

1860: *Lois fondamentales de la genèse spontanée...*

1860: *Corps organisés recueillis dans l'air par les flocons de neige°*

1860: *Genèse des proto-organismes dans l'air calciné... à 150 degrés°*

1860: *Moyen de rassembler sur un très petit espace tous les corpuscules...°*

1860: *Analyse microscopique de l'air atmosphérique en différents lieux...°*

1861: *Générations spontanées. État de la question en 1860*

1861: *De la nature et de la genèse de la levûre dans la fermentation alcoolique°*

1861: *Des phénomènes biologiques des fermentations...°*

1862: *Les créations successives et les soulèvements du globe*

1862: *Recherches sur la résistance vitale de l'invisible...°*

1862: *Expériences sur les migrations des Entozoaires°*

1862: *Générations spontanées. Remarques critiques sur le mémoire du Vte d'Auvray*

1863: *Études expérimentales sur la genèse spontanée*

1863: *Générations spontanées. Résumé des travaux physiologiques sur cette question*

1863: *Observations sur l'air de la cîme du Mont-Blanc à 14800 pieds°*

1863: *Expériences sur l'hétérogénie exécutées dans les glaciers de la Maladetta°*

1863: *Des limites de la résistance vitale dans le vide°*

1863: *Réponse aux observations critiques de M. Pasteur...°*

1864: *Observations sur la neige de la cîme du Mont-Blanc...°*

1864: *Observations sur la prétendue scissiparité de quelques Microzoaires°*

1864: *Embryogénie des Infusoires ciliés°*

1864: *Mémoire sur l'embryogénie des Infusoires°*

1864: *Observations sur le développement des Infusoires ciliés°*

1864: *Production des Bactéries et des Vibrions dans les phlegmasies des bronches°*

1864: *Nouvelles expériences sur la génération spontanée et la résistance vitale* [271 p.]

Appendix 2

The Primary Archives on the Pasteur-Pouchet Debate

Rouen, Museum of Natural History

MHN, FAP 18 "Heterogeny-Miscellaneous" (notebooks, reports)
MHN, FAP 19 "Heterogeny-Miscellaneous" (notebooks, reports)
MHN, FAP 22 "Heterogeny-Collection" (conferences of Joly, Musset, etc.)
MHN, FAP 30 "Pouchet's Scientific Correspondance" [A–F]
        3012 Bernard
        3014 Blainville
        3047 Coste
        3066 Duruy (Minister of Education)
        3078 Flourens
MHN, FAP 31 "Pouchet's Scientific Correspondance" [G–Z]
        3102 Goeffroy Saint-Hilaire
        3176 Pasteur
        3192 Rayer
MHN, FAP 51–52 "Copies of Letters from Pouchet"
        (1853–1860, 1862–1872)

Rouen, Municipal Library

MS. P. 55 "Norman Autograph Manuscripts"
        Letters from Pouchet to various naturalists
MS. G. 182 (2) "Papers Concerning Pouchet"
        Letters from Flourens (4), Geoffroy Saint-Hilaire (1), and Blainville (1)
N° inv. 1754 "Don Levillain" (uncatalogued material)
        Letters from Pouchet to Coste, Noël, Moigno, etc.

Paris, Bibliothèque nationale de France

BnF, naf 18106 "Pasteur's Correspondance" (already published)
> 45–46: Pouchet's first letter to Pasteur (February 22, 1859)

BnF, naf 18111 "Letters received by Flourens (sic)" (misfiled)
> 4–341: Pouchet's Letters to Joly/Musset
> 363: Pouchet's Notes to the Institut
> 369: Pouchet's Notes to the Commission of the spontaneous generation
> 381–483: Joly's Letters to Pouchet
> 484–5: Lettres from the Institut to Joly
> 486-524: *Varia*

Appendix 3

Excerpts from Pouchet's correspondence

Note: Félix-Archimède Pouchet is hereafter denoted as FAP.

[1] Flourens to FAP, February 4, 1847, Rouen Museum, FAP 3078.
« Monsieur,
Je viens de recevoir votre excellent ouvrage, et votre bel atlas, et je m'empresse de vous en remercier. Je remettrai mardi à l'Académie l'exemplaire qui lui est destiné, et je tâcherai de le faire en des termes qui répondent à mon estime pour le livre et pour l'auteur. Je n'ai pas besoin de vous dire combien je suis touché de votre dédicace; je le suis surtout du noble sentiment vous l'a dictée. Je suis aussi très satisfait de ce que vous voulez bien m'accorder dans les progrès d'une science qui vous doit tant [...] »

[2] Flourens to FAP, October 3, 1848, Museum de Rouen, FAP 3078.
« Monsieur,
[...] Il y a deux jours que j'eus occasion de parler dans une de mes leçons de votre bel ouvrage sur l'*ovulation*; je me vantai devant mon auditoire de l'honneur que vous m'avez fait en me le dédiant: "honneur sérieux, ajoutai-je, car c'est-là un de ces ouvrages qui font époque de la science et qui y restent." Agréez tous mes remerciments et l'expression bien sincère de mon entier dévoûment. »

[3] Flourens to FAP, April 22, [1849], Rouen Museum, FAP 3078.
« Monsieur et cher confrère,
[...] Quant à votre lettre, croyez-bien que je n'oublierai pas votre bel ouvrage; je vais m'en occuper, et j'en parlerai à M. Vitet. Comptez sur le véritable attachement de votre dévoué confrère. »

[4] Flourens to FAP, April 17, 1853, Rouen Museum, FAP 3078.
« *Le secrétaire perpétuel de l'Académie des Sciences* à Monsieur Pouchet, correspondant de l'Institut, &c. Monsieur,

[Flourens congratulates Pouchet for his *Histoire des sciences naturelles au Moyen Âge*] [...] Cet ouvrage a été déposé dans la Bibliothèque de l'Institut. Agréez, Monsieur, l'assurance de ma [de la main de Flourens: considération très distinguée et de toute mon affection] »

[5] Coste to FAP, June 17, 1854, Rouen Museum, FAP 3047.
« Mon cher confrère,
[...] Avant de finir, je désire vous dire un mot d'un projet auquel j'ai souvent pensé. Vous êtes placé à côté d'un grand cours d'eau, non loin d'un littoral où il serait facile d'organiser des expériences sur la reproduction des espèces marines. Pourquoi ne profiterait-on pas de cette double position pour fonder deux établissements dont vous pourriez diriger les opérations, l'un sur les bords de la Seine, l'autre sur les bords de la mer? Je serais d'autant plus disposé à vous seconder dans cette entreprise, que ce serait pour moi un nouveau champ d'observation où j'aurais grand plaisir à être votre collaborateur [...] Croyez à mes plus vives sympathies. »

[6] Geoffroy Saint-Hilaire to FAP, March 2, 1855, Rouen Museum, FAP 3102.
« Cher collègue et ami,
Avant-hier le prince Ch. Bonaparte est venu me voir, et c'est lui qui m'a mis au courant de ce qui se passe (moins toutefois la composition du programme!). Heureusement il m'a mis aussi au courant du résultat de sa démarche qu'il a aussitôt faite pour vous, avec le plus amical empressement comme avec la plus parfaite conviction. Il vous aura sans doute écrit. Il croit la chose en aussi bon état que possible. Je crois donc que vous pouvez être tranquille, et je suis heureux, s'il y a lieu (et je vais me tenir au courant), d'ajouter mes faibles efforts aux siens [...] À vous de coeur. »

[6*] Pasteur to FAP, February 28, 1859, *Correspondance de Pasteur* II, 45.
« Monsieur,
J'ai reçu la lettre que vous avez bien voulu m'écrire à l'occasion de la note présentée en mon nom à l'Académie par M. Dumas, dans la séance du 14 février. Vous me faites beaucoup d'honneur, Monsieur, en paraissant tenir à mon avis sur la question de la génération spontanée. Les expériences que j'ai faites à son sujet sont trop peu nombreuses et, je dois le dire, trop changeantes dans les résultats

qu'elles m'ont offerts pour que j'ose avoir une opinion digne de vous être communiquée. [...]

Mais lorsque la liqueur est prête, faites-la bouillir quelques minutes dans un ballon effilé communiquant par un caoutchouc à un petit tube de cuivre qui est entouré de charbons ardents, puis laissez refroidir et alors fermez par un trait de chalumeau la partie effilée, puis portez le ballon dans une étuve. Il pourra y demeurer des mois entiers à une température de 25 à 35° sans donner aucune apparence de fermentation, ni levures, ni infusoires.

Veuillez, Monsieur, adopter la disposition que je vous indique; en moins d'un quart d'heure, vous pourrez mettre une expérience en train, vous acquerrez alors la conviction que dans vos expériences récentes, vous avez à votre insu introduit de l'air commun et que les conséquences auxquelles vous êtes arrivé ne sont pas fondés sur des faits d'une exactitude irréprochable. Je pense donc, Monsieur, que vous vous avez tort, non de croire à la génération spontanée, car il est difficile dans une pareille question de n'avoir pas une idée préconçue, mais d'affirmer la génération spontanée. [...]

Veuillez agréer, Monsieur, l'expression de mes sentiments les plus distingués. »

[7] Geoffroy Saint-Hilaire? to FAP, August 31, 1859, Rouen Museum, FAP 3102.

This is the letter allegedly sent by Isidore Geoffroy Saint-Hilaire to Pouchet. It has been shown in Chapter 2 that this letter was a fake written by Pouchet himself. The most dubious passages are underlined.

« Mon cher et très-honoré Confrère,

J'ai éprouvé le plus grand regret de m'être trouvé absent lorsque vous avez eu l'extrême amabilité de m'apporter vous-même l'ouvrage dans lequel vous reprenez, de fond en comble, la grande question de l'hétérogénie. Quoique Rouen et Paris se touchent presque, aujourd'hui, les occasions de vous voir sont rares: chacun de nous, retenu dans son cabinet, y vit pour la science, et lui sacrifie tout, même jusqu'aux jouissances des relations amicales et confraternelles.

Au moins aurais-je voulu vous remercier aussitôt par écrit. Mais j'étais accablé d'épreuves, dont la correction suspendait un départ arrêté pour le 25, et qui n'est pas encore effectué. Une des compensations que j'aurais éprouvées, pour la perte d'une semaine de mes très-courtes vacances, est le plaisir de vous lire. C'est à votre livre [2] que j'ai donné tous les moments que je pouvais enlever à mes travaux, rendus urgents par la nécessité de publier mon demi volume avant mon départ, ou du moins de l'avancer tellement que je n'aie plus à m'occuper à mon retour que de la suite.

C'est ce qui fait que je suis encore peu avancé dans la lecture de votre ouvrage. Je viens de terminer ce matin la métaphysique. J'avais lu hier la partie historique.

Je crois qu'on ne pouvait mieux poser, que par <u>ces deux articles si savants et si philosophiques,</u> la question que vous traitez; ce sont <u>les préliminaires les plus propres à la poser dans toute sa grandeur,</u> et à la dégager de tous les éléments étrangers dont on l'a compliquée, au moins autant qu'il est possible, car vous ne parviendrez jamais à empêcher qu'une foule de personnes (qui ne sont pas toujours les plus religieuses en réalité!) vous opposent les arguments théologiques, malgré la *démonstration* que vous donnez de la liberté laissée ici à la science [3] par la théologie. — <u>La science a d'ailleurs ses voies propres, et elle a le droit d'y marcher sans se préoccuper d'autre chose que de la recherche de la vérité.</u> — Mais on ne le comprend pas encore assez, et il y a, dans le choix du sujet que vous avez pris, <u>autant de courage</u> à ce point de vue, qu'il vous en faut d'ailleurs pour lutter contre les difficultés propres du sujet.

Vos appréciations scientifiques m'ont paru parfaitement justes. Vous jugez bien vos devanciers, vous faites bien ressortir en quoi ils ont avancé, arrêté ou <u>fait dévier la science.</u> J'ai vu en particulier, avec une grande satisfaction, <u>la justice que vous rendez à Lamarck, si injustement traité</u> durant sa vie et après sa mort: je me propose de faire parvenir ce que vous dites à la connaissance de son fils, inspecteur général des Ponts et Chaussées, qui est venu souvent à mes cours, et m'a adressé souvent des lettres relatives, soit à mes sympathies pour les travaux de son père, soit à mes dissidences sur divers points. [4] Le seul de vos devanciers envers lequel je crois qu'il y aurait un peu à réclamer, c'est Bonnet. Je suis loin d'en partager les vues générales; mais que de belles et ingénieuses idées émises à l'appui d'erreurs que je n'appellerai pas belles, car <u>il n'y a de beau que le vrai</u>, mais qui sont d'ingénieuses et brillantes conceptions dont peu d'esprits sont capables.

Je reprendrai ce soir & demain matin votre livre. J'ai hâte d'arriver à vos expériences. <u>Je n'ai pas ici de parti pris.</u> Je crois ce que vous me dites de <u>la puissance créatrice continuée dans le temps, très sage, très élevé et, dans le vrai sens, très religieuse.</u> La génération spontanée est, comme vous le dites très-bien, rejetée par la plupart sur l'autorité du maître. Or, en science, l'autorité du maître est un pauvre argument; car l'élève sait ce qu'a su le maître, et nécessairement quelque chose de plus. Venons donc au fait, à l'expérience; ce sont là les *vrais maîtres.* Aussi ai-je beaucoup contribué à faire admettre *la question de prix proposée par l'Académie et dont le choix a dû vous satisfaire et vous flatter; car, avant vos communications à l'Académie, qui eût pensé, qui eût osé lui proposer un tel sujet?* C'est à M. Flourens que revient le mérite de l'avoir mis en avant; mais je puis *me rendre cette justice* que son adoption m'est due; car nos trois autres collègues, par conséquent la majorité, ne voulaient pas qu'on parût *douter,* et M. Flourens qui n'aime pas beaucoup la lutte, allait laisser tomber sa proposition. Seulement il a fallu concéder une rédaction qui n'engageât nullement l'Académie; et ici, il y avait d'ailleurs sagesse à ne pas s'engager. *Cette question, en effet, a des profondeurs immenses, et plus on y pénètre, et plus on s'en convainc.* La distinction que vous faites à la fin de votre partie métaphysique me semble difficile elle-même à admettre d'une manière absolue. Temporairement, pour le présent, et pour longtemps encore, oui. Mais ensuite? Qui le sait. Vous entreprenez de démontrer qu'avec des *éléments organiques,* on peut créer des êtres organisés, et il est certain [6] qu'il y aurait folie à vouloir aller au-delà aujourd'hui. Mais pourra-t-on arriver un jour à créer des éléments organiques, ou pour parler ici la langue chimique, des principes immédiats de toute pièce? Vous savez ce qu'ont déjà fait dans cette direction quelques chimistes, principalement Mr Berthelot et vous semblez traiter la première et la seconde partie de la même question, et si vous avez tous deux raison, il s'en faudra de bien peu que Lamarck n'ait eu raison d'être aussi audacieux.

En voilà bien long, cher Collègue et ami, et cependant je ne terminerai pas sans vous dire combien je suis reconnaissant de la belle justice que vous rendez à mon père, et vos trop bienveillantes paroles envers moi. J'y vois une preuve de votre amitié plus que de votre justice, et si elles me sont moins flatteuses, ainsi comprises, elles m'en sont plus chères. – Je veux vous dire aussi que j'ai lu votre préface avec un sentiment de sympathie bien [7] sincère; vous y parlez des uns pour les remercier; vous vous taisez sur d'autres dont vous aviez le droit de vous plaindre: cette modération vous honore, et en vous élevant d'autant plus au-dessus de telle injustice, que nous n'avons pas oublié, vous montrez d'autant mieux combien elle a été grande et reste injustifiable.

Adieu, agréez l'expression de mes sentiments très-distingués et dévoués.

I. Geoffroy St Hilaire. »

[8] FAP to Pasteur, February 22, 1859, BnF, MS. naf 18106, fol. 45.

« Monsieur et très honoré confrère,

J'ai vu dans un journal scientifique que votre expérience était un argument *contre* les générations spontanées. Je viens de faire lire les comptes rendus à plusieurs savants et ils me disent que vous êtes *pour* les générations spontanées [...] Trop délicat pour vous prêter des vues qui ne seraient pas les vôtres, veuillez, je vous prie, dissiper les doutes qui se traduisent dans l'opinion. C'est en trop grand nombre que sont les preuves de g.[énération] sp.[ontanée] pour que j'ai besoin d'en trouver de nouvelles [...] »

[9] FAP to Noël, January 2, 1860, B.m. Rouen, MS. 1754 inv.

« Vaillant et sublime ami,

Merci encore de votre splendide souhait pour l'hétérogénie. Je puis vous assurer qu'elle va triompher, dans un temps prochain, de tous les obstacles. Chaque jour je renverse un peu de l'édifice des panspermistes et je l'ai tellement miné qu'il ne tient plus qu'à un fil [...] »

[10] FAP to Noël, June 19, 1860, B.m. Rouen, MS. 1754 inv.

« *Homo sapiens*,

Les hommes de génie comme vous et moi, pour parler avec la plus stricte modestie, s'entendent toujours à merveille; aussi en lisant les premières lignes de votre aimable mémoire, j'en rêvais déjà la conclusion [...] »

[11] FAP to Joly, July 17, 1860, BnF, MS. naf 18111, fol. 8 r°.

« [...] Je vais continuer, car, selon moi, ce savant chimiste en sortant de son domaine, s'est évidemment égaré. Je n'ai pas été le chercher, il s'est de lui-même fourvoyé dans une question qui n'est point une fermentation ni une opération de son ressort mais bien une haute question du notre [footnote: Je n'ai plus qu'à lui prouver qu'une chose: que l'eau bouillante fait cuire les oeufs] [...] »

[12] FAP to Joly, October 24, 1860, BnF, MS. naf 18111, fol. 13 v°.

« [...] Oui, l'article [a newspaper article ordered by Pouchet] est sévère, mais pourquoi M. Pasteur s'avance-t-il dans une voie qui n'est pas la sienne et pourquoi s'y égare-t-il d'une si extraordinaire manière? [...] »

[13] FAP to Joly, November 4, 1860, BnF, MS. naf 18111, fol. 17 r°-18 v°.

« [...] Je viens de mettre la dernière main à ce que j'appelle mon *document magistral* [...] Je crois qu'elle [cette noble cause] est de nature, par son grand retentissement, à jeter *plus de gloire sur nos noms* que tout autre sujet [...] Vous verrez, si déjà vous ne l'avez pressenti, que le savant chimiste s'est suicidé [...] »

[14] FAP to Joly, November 11, 1860, BnF, MS. naf 18111, fol. 19 r°-19 v°.

« [...] Il est vrai que pour venir en aide à ses expériences, il lui *a fallu inventer des animaux et des plantes qui vivent dans l'eau bouillante!!!!!* Tout cela est vraiment merveilleux, féérique même [...] Le malheureux ne s'aperçoit donc pas que *son invention des animaux qui vivent dans l'eau bouillante suicide toutes ses expériences à l'avance* [...] »

[15] FAP to Joly, mid-November 1860, BnF, MS. naf 18111, fol. 21 v°.

« [...] Je le combats à la face du ciel pour notre cause commune [...] Au sujet de cette *sainte cause*, à laquelle vous donnez un si magnifique appui, *ne craignez nullement de prendre les devants* — courage, donc, et *marchez en avant, tant que nous aurons la convicition d'être dans le vrai* [...] »

[16] Geoffroy Saint-Hilaire to FAP, March 3, 1861, Rouen Museum, FAP 3102.

« [...] L'objet de cette lettre est de vous faire une communication, au nom de la Commission des conférences établies dans le sein de la Société d'acclimatation [dont Pouchet est membre fondateur], et qui obtiennent du succès. Tout le monde trouve que c'est une excellente constitution. Les Ministres nous en osent féliciter; le prince Napoléon ne veut pas s'en tenir là; c'est par sa présence qu'il veut témoigner de l'intérêt qu'il prend à cette nouvelle expérience de notre oeuvre [...] Mais plus nos conférences intéressent, et plus il faut en relever l'intérêt. C'est pourquoi la Commission m'a chargé de vous écrire et de vous demander si vous ne pourriez, prévoyant un voyage à Paris en mai ou en avril, nous donner une heure [...] Vous choisirez, en cas d'affirmative, votre jour, votre heure, votre sujet. Ils sont acceptés d'avance. Votre très dévoué. »

[17] Geoffroy Saint-Hilaire to FAP, March 18, 1861, Rouen Museum, FAP 3102.

« Cher confrère et ami,

[...] Quant à votre travail *État de la question* [this is Pouchet's *Générations spontanées* of 1860] vous me l'avez déjà envoyé, et c'est pourquoi je puis immédiatement, non seulement vous remercier, mais vous féliciter. C'est bien, comme vous le dites, un résultat commencé que d'avoir ébranlé l'opinion, et c'en est un autre que d'avoir déjà éliminé tant de prétendus résultats d'expériences qu'on tenait pour la vérité même. Nous devons aussi, nous naturalistes, vous savoir ce gré infini d'avoir revendiqué les droits de notre science, et de signaler ce qu'il y a de faux à prétendre résoudre par des expériences de chimiste des questions de cet ordre. La chimie peut y être la très utile auxiliaire de la biologie, mais la question reste essentiellement biologique, et il faut, pour la bien traiter, y apporter l'esprit biologique. Courage donc, et continuez à soutenir haut et ferme le drapeau de notre science, inconnue aux chimistes et aux physiciens, et où il semble cependant qu'ils aient le droit de venir nous faire la leçon. Votre bien dévoué confrère. »

[17*] Pasteur to FAP, April 3, 1861, *Correspondance de Pasteur* II, p. 84.

« Monsieur et Cher Confrère,

La différence de nos opinions sur la question célèbre des générations spontanées ne m'empêche pas d'estimer très haut votre labeur et les louables efforts que vous faites pour introduire l'expérience dans les questions d'histoire naturelle. [...]

Je lis avec beaucoup de soin tout ce que vous écrivez sur le sujet qui nous occupe tous deux. Or, je ne puis pas me procurer une brochure que vous venez de publier, me dit-on, et qui serait la reproduction d'articles insérés dans un journal de médecine que je ne reçois pas.

Je serais heureux d'en avoir un exemplaire parce que je rédige en ce moment l'ensemble de mes observations, où tout naturellement je critique vos assertions. J'espère aussi faire prochainement un résumé de mon travail dans une des leçons de la *Société chimique*.

Veuillez agréer, Monsieur et Cher Confrère, avec l'assurance de ma parfaite estime, l'hommage de mes sentiments respectueux. »

[18] FAP to Pasteur, April 6, 1861, BnF, MS. naf 18106, fol. 47–48.

« Cher savant et confrère,

Si vous n'avez point encore reçu ma brochure, c'est que je me proposais de ne vous l'envoyer qu'en ayant l'honneur de vous écrire, et le temps m'a manqué. Je vous l'adresse aujourd'hui même par la poste. Si, dans cet écrit, j'attaque vivement le physiologiste, je rends hommage partout au célèbre chimiste, dont les travaux excitent mon admiration. J'ai voulu revendiquer pour ma science un sujet qui est essentiellement de son domaine, et qui plane même dans ses plus hautes régions. Il s'agit simplement ici d'une question d'ovologie microscopique. La chimie peut lui venir accessoirement en aide, mais elle n'en appartient pas moins essentiellement à la biologie. Je vois et je ferai voir à tout Paris l'ovule des oxytriques se formant de toutes pièces. Divers observateurs l'ont vu *avant moi*, d'autres l'ont *vu après*; je vois se former de la levûre *où* je constate qu'il n'en existait pas. *Je la vois germer et fructifier*; et, dans mon laboratoire, déjà beaucoup de personnes ont vérifié ce que j'avance [note de bas de page: même plusieurs chimistes! (Houzeau?)] La chimie, quelque puissante qu'elle soit, *jamais* n'anéantira de tels faits [...] Mais combien vous êtes déjà loin, mon cher confrère, des opinions que vous émettez dans ce mémoire et dans celui que vous professez aujourd'hui. Il y a la distance qui existe entre le jour où vous m'écriviez cette ligne *"il n'y a pas plus de germes organisés dans l'air qu'il n'y a de germes de sulfate de soude"*. Vous avez imité M. Dumas. L'illustre chimiste est loin aussi du temps où son oeil exercé voyait toutes les molécules se grouper *spontanément* pour former un organisme! [...] »

[18*] Pasteur to FAP, April 8, 1861, *Correspondance de Pasteur* II, p. 85.

« Monsieur et cher Confrère,

Je vous remercie de l'empressement que vous avez mis à m'envoyer la brochure que je vous avais demandée. Que de choses j'aurais à y répondre! Mais ne parlons, si vous le voulez, que de votre aimable lettre. Vous voyez, dites-vous, la levure de bière se former spontanément et germer, et fructifier... tout ce qu'affirme, en un mot, votre récente communication à l'Académie des sciences.

Eh bien, mon cher Confrère, il est démontré pour moi qu'il n'y a pas une de vos assertions sur la levure qui ait la moindre exactitude. Tout ce que l'on a écrit avant vous dans le même sens n'a pas plus de valeur. Si vous pouvez me montrer de la levure de bière donnant un *Aspergillus* je vous rends les armes. [...]

Prise à sa date, je n'ai pas un mot à retrancher à ma lettre de 1859 [6*] dont j'ai la minute exacte sous les yeux.

Veuillez agréer, mon cher Confrère, l'assurance de mes sentiments de haute et parfaite considération. »

[19] FAP to Pasteur, April 9, 1861, BnF, MS. naf 18106, fol. 49–50.

« Mon cher confrère,

Je relis la lettre que vous m'avez fait l'honneur de m'écrire et dont je vous adresse la copie textuelle. Je l'ai alors bien mal interprétée, si ce que j'y vois ne signifie pas, sommairement, qu'il n'y a pas plus de germes organisés dans l'air qu'il n'y a de germes de sulfate de soude... mais passons. Si vous consentiez à me rendre les armes, pour me servir de votre expression, cher et savant confrère, si je vous montre de la levûre germant et fournissant des *Aspergillus*, le combat finirait à l'instant. Quand il s'agit d'observations délicates de physiologie, je n'affirme les choses qu'après le plus sérieux examen, et il est rare qu'on puisse me convaincre d'erreur [...] Au mois d'août prochain, j'irai m'installer au Jardin des Plantes et je puis vous assurer que je montrerai à qui le voudra de la levûre naissant spontanément,|de la levûre germant, de la levûre commençant à pousser des tiges, de la levûre en fructification,|et je vous assure aussi, cher et savant confrère, que comme il s'agit là de faits de biologie bien vus, qu'on produit facilement quand on sait s'y prendre, toutes les expériences chimiques du monde, je le répète, ne renverseront pas cela. Mais, cher et savant confrère, c'est qu'il me semble que, chimiste d'une célébrité incontestable, quand il s'agit de biologie vous confondez quelques fois les choses, permettez-moi de vous le dire. Vous prétendez "que mes assertions sur la *levûre de bière* n'ont pas le moindre fondement—que tout ce que l'on a écrit dans le même sens n'a pas plus de valeur..." C'est là un exemple de ce que je viens d'avoir l'honneur de vous exprimer. Je n'ai pas dit un mot de la levûre de bière!... Je ne me suis encore occupé que de la *levûre du cidre*, dont personne n'a encore parlé, que je sache... Pour la levûre de bière, je la vois germer ainsi que d'autres savants qui s'en occupent aussi en ce moment; sous peu je vous dirai, je l'espère, quels sont les végétaux qu'elle produit. Les auteurs qui n'ont parlé que de cette levûre, Pelouze et Frémy, semblent avec beaucoup d'autres être certains qu'elle produit du *Penicillium* [...] Au moment où je termine cette lettre, j'ai dans le champ de mon microscope des semences de l'*Ascophora mucedo*, Tad. si commun dans nos expériences; et je les fait voir à un professeur d'histoire naturelle de l'une de nos facultés... *C'est apparent comme le soleil*, me dit-il [...] Vous ne trouverez pas de spores de levûre dans aucune des parcelles de l'atmosphère qui me serviront à produire des fermentations! [...] »

[20] Geoffroy Saint-Hilaire à FAP, September 10, 1861, Museum de Rouen, FAP 3102.

« Mon cher collègue et ami,

[...] Bien sensible à vos amicales félicitations, j'éprouve un grand regret de ne pouvoir vous envoyer les miennes en échange. Mais l'acte de justice que j'espérais, ne peut être que différé, et de peu de temps. Il en sera en cette occasion comme l'an dernier de M. Duméril [...] Votre très dévoué. »

[21] FAP to Joly, November 8, 1861, BnF, MS. naf 18111, fol. 62 v°.

« [...] Mettez donc votre mémoire côte à côte avec l'une des élucubrations embrouillées de M. Pasteur, et vous reconnaîtrez que je suis dans le vrai en réclamant pour l'hétérogénie la palme de la dialectique [...] »

[22] FAP to Joly, February 2, 1862, BnF, MS. naf 18111, fol. 71 r°-72 v°.

« [...] M. Milne-Edwards règne dans la section de zoologie; et règne absolument depuis la mort de Geoffroy St-Hilaire. MM. de Quatrefages et Valenciennes suivent volontiers ses impulsions; aussi je désespère absolument [...] De quoi vous plaignez-vous, sapristi! On parle de nous sur les déserts rivages de l'Afrique — je ne ris pas. On s'y réjouit à Port-Saïd quand vous ou moi rossons M. Pasteur [...] »

[23] FAP to Joly, February 8, 1862, BnF, MS. naf 18111, fol. 73 r°-75 r°.

« [...] Vous me demandez si, sapristi, vous n'avez pas été trop loin — Non, sur ma foi [...] Ne perdez pas un instant, écrivez au Ministre [...] Dans l'intérêt de votre candidature à la Sorbonne, *ne perdez pas un instant*, remuez tout [...] »

[24] FAP to Joly, March 4, 1862, BnF, MS. naf 18111, fol. 78 r°.

« [...] M. Roulland, auquel je me suis adressé, comme vous le savez, m'a fait avoir raison de l'infâmie de M. Milne-Edwards. Dans son discours, inséré dans la *Revue ministérielle* des Sociétés savantes, il y a une rectification à mon égard [...] »

[25] FAP to Noël, March 26, 1862, B.m. Rouen, MS. 1754 inv.

« Cher et savant ami,

Je viens de lire votre admirable article, et je ne veux pas laisser passer un seul instant sans vous en remercier bien cordialement [...] Il ne pèche que par la manière bienveillante dont vous vous exprimez à l'égard du savant... [i.e. Pasteur] »

[26] FAP to Joly, April 9, 1862, BnF, MS. naf 18111, fol. 84 r°-85 v°.

« [...] Si M. Roulland eut pris une vigoureuse initiative, mais vigoureuse, vous eussiez déjà pu l'emporter [...] Si vous voulez m'en croire, tombons sur le dos de ces *vieilleries chimiques* [...] »

[27] FAP to Noël, no date, B.m. Rouen, MS. 1754 inv.

« Mon cher ami,

Vous êtes vraiment donc d'une bonté biblique, monastique, angélique et séraphique [...] Voici deux de nos grande sommités qui croient aux miracles d'un farceur ou d'un illuminé [...] [The latter:] M. Roulland, ancien chef des libres penseurs de Rouen, voltairien de souche, qui, aujourd'hui, n'a pas assez d'encens à brûler au nez des cardinaux [...] »

[28] FAP to Joly, April 9, 1862, BnF, MS. naf 18111, fol. 112 r°-113 r°.

« [...] J'étais venu à Paris avec des instruments pour faire quelques expériences avec M. Flourens. C'est peine perdue. Il est absent et malade [...] M. Serres, le seul que j'ai pu rencontrer, me paraît nullement hostile à nos travaux et plutôt disposé à les appuyer [...] »

[29] FAP to the Abbé Moigno, October 4, 1862, Rouen Museum, FAP 52.

« Très Savant Abbé,

[...] Je vous remercie beaucoup de la gracieuse manière dont vous avez annoncé l'arrivée de mon manuscrit à l'Académie. Mais pourquoi, savant Abbé, toujours protester contre des phénomènes qui, *bien étudiés*, sont aussi évidents que la lumière. Laissez donc toutes ces expériences chimiques qui sont ridicules et dans lesquelles la vanité de leurs auteurs prétend se substituer à Dieu; laissez la nature agir et tout va se dérouler à vos yeux. J'ai fait un monceau de ruines des expériences si vantées de Schulze et de Schwann, et de toutes les contradictions de M. Pasteur qui n'en est que le continuateur moins précis, moins exact, avec son vide de la machine pneumatique [...] Très savant Abbé, quand Paysonnel, Goethe, Ehrenberg, etc., ont présenté leurs travaux aux savants français, il leur a fallu dix ans de luttes pour leur faire accepter tant de vérités. Si on m'y force, je resterai autant qu'eux sur la brêche [...] [Note: Vous allez me voir un de ces jours tuer définitivement les résurrections que j'ai tant ébranlées [...] »

[30] FAP to Joly, October 5, 1862, BnF, MS. naf 18111, fol. 123 v°.

« [...] Je ne cesserai jamais la bataille; *j'en jure bien par le Styx et les dieux infernaux* [...] »

[31] FAP to Joly, October 13, 1862, BnF, MS. naf 18111, fol. 130 r°-131 r°.

« [...] Le prix ne sera décerné à personne dit M. Blanchard. Je le pense aussi actuellement [...] Ainsi nous avons: MM.— M. Edwards contre, Bernard contre, Flourens contre, Brongniard? Coste? »

[32] FAP to Joly, October 21, 1862, BnF, MS. naf 18111, fol. 132 r°.

« J'ai revu mon laboratoire, noble et cher ami, et pour défendre notre sainte cause, je vais y arborer l'oriflamme. Vous ne lâcherez pas les pieds, dites-vous! Mais moi non plus! [...] Monsieur Pasteur nous a traité d'*ignorants* dans ses leçons au cercle chimique. Il paiera l'affront d'une sanglante manière. Comme je sens qu'auprès de lui nous avons la force d'Antée, je ne l'abandonnerai qu'étouffé sous le poids des rochers de l'hétérogénie [...] Occupons-nous de M. Flamel [i.e. Pasteur] pour l'écraser [...] »

[33] FAP to Joly, November 12, 1862, BnF, MS. naf 18111, fol. 140 r°-140 v°.

« [Pouchet is coming back from his journey to Paris, where he enquired about the voting intentions of the members of the commitee] Je viens de faire un nouveau pélerinage dans cette affreuse Babylone. *Trois fois abomination* [...]

[34] FAP to Joly, January 19, 1863, BnF, MS. naf 18111, fol. 154 r°-159 r°.

« [...] Ne vous y trompez point, nous avons, quoique battus, remporté une *grande victoire* [...] Dans la lutte à mort qui va s'engager, vous ne me trouverez pas moins courageux que vous [...] Votre origine et la mienne sont un ineffaçable péché [...] ».

[35] An auditor to Joly, March 2, 1863, BnF, MS. naf 18111, fol. 490 r°.
« Monsieur,
Converti à votre croyance en la génération spontanée par l'exposé si empreint de conviction de vos expériences et par le prestige de votre parole [...] »

[36] FAP to a Colleague, March 18, 1863, B.m. Rouen, MS. p. 55.
« [This letter comes along with a copy of *Hétérogénie*, the preface and the introduction of which Pouchet requests to be read] Dans l'une, vous verrez que je combats avec la plus belle armée qu'on puisse imaginer. Dans l'autre, que si dans nos provinces tout le monde avait cette noblesse de courage qui vous anime, et que je tache d'imiter, les hommes de lettres et les savants qui s'y trouvent éparpillés seraient un peu plus considérés et honorés; et l'on ne verrait pas le gouvernement leur adresser que de rares et dérisoires encouragements, tandis qu'il accable de toutes ses faveurs le seul petit coin de Paris [...] [About provincial savants:] Les uns sont morts [marginal annotation: *negotium infandum*] sans obtenir l'une de ces distinctions dont M. le Ministre *honore* tant de ses nullités universitaires; D'autres, que je m'honore de compter parmi mes amis auront peut-être le même sort? »

[37] FAP to Noël, no date [May 1863?], B.m. Rouen, MS. 1754 inv.
« Mon cher ami,
Je m'empresse de vous apprendre que l'hétérogénie vient pour la seconde fois de mettre l'Institut à feu et à sang. Tohu bohu général. Quelques personnes ont proposé de m'appeler / immédiatement / à Paris pour voir les expériences de M. Pasteur — de *prudents amis* ont demandé un sursis. Amen [...] Pour l'article du Journal de Rouen, vous m'obligeriez *beaucoup* s'il peut être publié rapidement [annotations in red ink: Je vous dirai pourquoi confidentiellement] »

[38] FAP to the Minister of Education, May 23, 1863, BnF, MS. naf 18111, fol. 184 r°.
« [Pouchet explains his absence from a meeting held at the Sorbonne: according to him, he never requested, either verbally or in writing, to take the floor. He goes on to cite his dispute with Pasteur and his invincible superiority over his detested rival:] En trois minutes, Mr Pasteur eut été mis dans l'impossibilité de répondre; parce qu'avec moi, qui connaît le sujet à fond, il le sait, il n'y a pas de subterfuges possibles [...]

En me combattant, Mr. Pasteur [...] a conquis une certaine célébrité [...] Devenu le champion d'une côterie scientifique [...] »

[38*] Pasteur to FAP, *Correspondance de Pasteur* II, p. 142.
« Monsieur et très honoré confrère,
J'ai lu la note que vous avez insérée aux Comptes rendus de la séance de l'Académie du 21 septembre dernier. Serais-je indiscret en vous priant de me faire savoir ce qui est arrivé aux ballons B, C, G, H, dont votre note ne parle pas. [...]

[38**] FAP to Joly, October 31, 1863, *Correspondance de Pasteur* II, p. 143.
« Noble et cher Ami,
Je reçois à l'instant cette lettre de M. Pasteur, à laquelle je vous prie de répondre vous-même. Ce sera mieux que provenant de moi.
Vous voyez où il veut en venir. Pondérez bien vos phrases... »

[39] Flourens to FAP, February 25, 1864, Rouen Museum, FAP 3078.
« *Le secrétaire perpétuel de l'Académie* à Monsieur Pouchet, correspondant de l'Académie des Sciences à Rouen. Monsieur,
Vous avez désiré que vos expériences pussent être répétées devant une commission de l'Académie. De son côté, M. Pasteur a bien voulu accéder à ce désir. Une commission a été nommée. Cette commission vous invite à vous trouver à Paris pour procéder à ces expériences qui auront lieu du 1er au 15 mars. Les frais qui pourraient être occasionnés par suite de votre déplacement, ou de vos expériences, seront à la charge de l'Académie. Veuillez agréer, Monsieur, l'assurance de ma considération distinguée. »

[39*] Pasteur to Colonel Favé, April 3, 1864, *Correspondance de Pasteur* II, p. 160.
« Monsieur,
Je m'empresse de vous remercie de l'envoi de la notice de vos travaux que j'ai lue avec un grand intérêt, malgré mon ignorance sur ces matières.
Que j'étais loin de prévoir, Monsieur, que si j'avais un jour la satisfaction d'apprendre que vous étiez candidat à une place vacante à l'Académie, je serais parmi ceux qui ne contribueraient pas à votre succès. C'est cependant le vif chagrin qui m'était réservé. Je ne vois pas le moyen de ne pas voter pour M. Foucault [...]

[40] Flourens to FAP, June 6, 1864, Rouen Museum, FAP 3078.

« *Le secrétaire perpétuel de l'Académie* à Monsieur Pouchet, correspondant de l'Académie des Sciences à Rouen. Monsieur,

J'ai l'honneur de vous rappeler l'engagement que vous avez pris de vous trouver à Paris le 15 juin pour la répétition des expériences relatives à la *génération spontanée*. Veuillez agréer, Monsieur, l'expression de ma considération très distinguée. »

[41] FAP to Noël, Friday [June 17, 1864], B.m. Rouen, MS. 1754 inv.

« [Following the meeting of the board of experts convened by the Academy to consider Pasteur's and Pouchet's experiments on spontaneous generation] L'expiation de Théophraste Bombast Paracelse [i.e. Pasteur] a commencé hier. Il était fort mal à son aise au milieu de nous [...] »

[42] FAP to Noël, Sunday [June 19, 1864], B.m. Rouen, MS. 1754 inv.

« Par compensation, je pense que l'opinion publique est fortement pour nous, et l'on répète partout que l'infâme leçon de Paracelse II [i.e. Pasteur] a fait un effet déplorable [...] »

[43] FAP to Noël, Wednesday [June 22 or 29, 1864], B.m. Rouen, MS. 1754 inv.

« [...] Veuillez donc dire à ceux de vos amis chargés de ce soir qu'on se hâte tandis que la génération spontanée fait ici *fanatisme* [...] »

[44] FAP to the Abbé Moigno, June 28, 1864, Rouen Museum, FAP 52.

« Très Savant Abbé,

[...] M. Milne-Edwards disait dans l'un de ses discours, au sujet de M. Boucher de Perthes, que le sol français était fécond en innovations, mais qu'elles ne s'y implantaient définitivement *qu'après avoir fait un stage à l'étranger*. Si j'étais Empereur de France, après avoir lu ces désespérantes lignes, j'aurais immédiatement fait appeler l'illustre académicien pour l'admonester vertement, et lui signifier qu'on n'ait *jamais*, dans mon empire, à redire de si accablantes choses! [...] »

[45] FAP to Noël, Thursday [June 30, 1864], B.m. Rouen, MS. 1754 inv.

« Cher et aimable ami,

J'ai été voir le Doyen. Notre oeuvre ne marche pas — Nous aviserons ensemble aussitôt mon retour [...] [Regarding Joly's lecture, dated

28 June 1864:] Plus de 4000 personnes ont essayé d'y pénétrer. Et quand Joly est sorti, aux applaudissements de toute la salle, la foule de la cour l'a reçu en criant: "Vive Joly, vive l'hétérogénie" [...] Lui-même [i.e. Pasteur] a failli passer un mauvais quart d'heure en sortant de la séance, et a été pourchassé aux cris de: "À bas les Jésuites" (ceci entre nous, mais c'est certain) [...] »

[46] FAP to Noël, July 3, 1864, B.m. Rouen, MS. 1754 inv.
« Pasteur a été écrasé par Joly [...] Si ce que j'ai fait cent fois ne me manque pas dans les mains, ce coup sera mortel pour lui [...] »

[46*] Pasteur to Bertin, July 3, 1864, *Correspondance de Pasteur* II, 1951, p. 168.
« Mon cher ami,
[...] Je sais trop à qui j'ai affaire, surtout après ce qui vient de se passer devant la commission de l'Académie. La reculade du mois de mars aurait dû faire prévoir celle du mois de juin. Cependant j'avoue que je n'y croyais pas. [...]
Cette reculade du mois de juin est selon moi un acte inqualifiable. Tu sais s'ils ont été affirmatifs quand je leur ai porté le défi de donner devant témoins la preuve expérimentale de leurs assertions: *Nous relevons le gant. Si un seul de nos vases demeure inaltéré, nous avouerons loyalement notre défaite.*
Est-ce clair? Tu te souviens du résultat auquel cette phrase se rapporte. Telle est l'expérience par laquelle la commission a voulu commencer, car il faut bien commencer. Et comment aurait-elle pu avoir la pensée de débuter par une autre que celle qui avait donné lieu au défi et à la nomination de cette commission? Et qui donc a appelé l'attention de l'Académie sur cette expérience, si ce n'est *eux*, lorsqu'ils ont écrit qu'au prix des plus grands obstacles ils étaient allés la répéter dans les glaciers de la Maladetta. [...]
Qu'ils opèrent comme ils voudront. Leur expérience est fausse et de nouveau je les mets au défi de la produire devant témoins avec le résultat publié par eux.
La commission est toute prête à faire répéter d'autres expériences. Elle ne s'est pas borné à le leur dire de vive voix. Elle le leur a écrit ainsi qu'à moi dès le 18 juin, et il faut être bien audacieux pour le nier et pour abuser ainsi le public. Tout ce que demande la commission et elle eût été insensée d'agir autrement, c'est de procéder expérience par expérience. [...]

Mais je le reconnais avec toi, il n'y avait rien à faire. Ils ne comprennent pas ce que c'est qu'une preuve. L'exactitude leur est inconnue. La preuve pour eux est une demi-probabilité. Ils n'ont pas de principes, mais seulement des opinions. Heureusement, comme l'a dit Fontenelle, le règne des mots et des choses occultes est passé depuis qu'il y a des Académies. [...] »

[47] FAP to Noël, Friday, July [8], 1864, B.m. Rouen, MS. 1754 inv.
« [One can foresee] une déroute complète de l'armée de Paracelse II. Ce sont vos amis [journalistes] qui les ont acculés jusqu'au point de revoir le grand cheval de bataille d'Harvey *omne vivum ex ovo*. C'est une vraie décadence [...] »

[48] FAP to Noël, July 17, 1864, B.m. Rouen, MS. 1754 inv.
« Cher et splendide ami,
Maintenant le triomphe unanime a calmé bien des inquiétudes [...] L'éloquence de Joly a passé sur le corps de Pasteur. Je viens de revoir aujourd'hui Coste, et lui ai reparlé de vous. Il me promet de vous faire métamorphoser non en taureau [...] mais en inspecteur, ce qui n'est que mieux [...] »

[49] FAP to Noël, no date, B.m. Rouen, MS. 1754 inv.
« [...] G. Flourens, que j'ai vu élever, que son père recommandait à mon affection, et qui loin d'être le farouche tribun dont on parle, était le plus doux enfant du monde, et ensuite, le plus charmant jeune homme qu'on puisse trouver: *quantum mutatus ab illo!* »

[50] Rayer to FAP, November 29, 1869, Rouen Museum, FAP 3192.
« Monsieur et savant confrère,
Je me suis empressé d'offrir à la Société de Biologie vos nouvelles études sur la génération spontanée. Je ne vous dissimulerai pas, qu'à cette occasion, un membre qui faisait partie de la Commission chergée d'examiner les points en discussion entre vous et feu M. Doyère, a manifesté quelque étonnement de ce que vous n'aviez pas tenu compte suffisant des conclusions de la Commission. Pour moi, je suis avec intérêt ces études si difficiles et si délicates et je m'empresse de vous adresser tous mes remerciements pour l'envoi que vous avez bien voulu me faire de vos nouveaux travaux. Veuillez agréer, Monsieur et savant confrère, la nouvelle assurance de ma plus haute considération et de mes sentiments dévoués. »

Appendix 4

The Works published by the supporters of vitalism and organicism

Note: These bibliographic records have been drawn from the *Catalogue général de la Bibliothèque nationale*. Where undated, the probable date of composition is mentionned. Reissues are cited between square brackets [...] and publications in periodical are preceded by an asterisk *.

Barthez, Paul-Joseph (1734–1806)

1761: *Quaestiones medicae duodecim... quas propungabit in Ludoviceo medico Monspeliensi...*

1773: *Oratio academica de principio vitali hominis...* [1863]

1774: *Nova doctrina de functionibus naturae humanae*

1778: *Nouveaux éléments de la science de l'homme* [1806]

1782: *Essai d'une nouvelle méchanique des mouvements progressifs* [1798]

1801: *Discours sur le génie d'Hippocrate*

1802: *Traité des maladies goutteuses* [1819, 1839]

1807: *Consultations de médecine de P.J. Barthez* [1810]

1816: *Mémoires sur le traitement des fluxions et sur les coliques iliaques*

1821: *Cours théorique et pratique de matière médicale-thérapeutique*

Dumas, Charles-Louis (1765–1813)

1787: *Mémoire couronné par la Société royale de médecine... sur la fièvre*

1790: *Quaestiones medicae duodecim propositae pro regia cathedra vacante...*

1797: *Système méthodique de nomenclature et de classification des muscles*

1800: *Principes de physiologie, ou Introduction à la science expéri-*
     mentale... [1806]
1803: *Discours sur les progrès futurs de la science de l'homme*
1812: *Doctrine générale des maladies chroniques* [1813]
1824: *Consultations et observations de médecine de feu*
     Ch.L. Dumas

Lordat, Jacques (1773–1870)

1813: *Conseils sur la manière d'étudier la physiologie de l'homme...*
1818: *Exposition de la doctrine de P.J. Barthez*
1833: *Deux leçons de physiologie faites à la faculté de médecine de*
     Montpellier
1833: *Essais sur l'iconologie médicale*
1837: *De la perpétuité de la médecine*
1844: *Preuves de l'insénescence du sens intime de l'homme*
1848: *De la nécessité de créer... une chaire de philosophie naturelle*
     inductive
1851: *Intentions didactiques de... l'Idée pittoresque de la physiologie*
     humaine
1852: *Accord de la doctrine anthropologique de Montpellier avec ce*
     que demandent les lois, la morale publique...
1853: *Théorie physiologique des passions humaines*
1854: *Réponses à des objections faites contre le principe du*
     dynamisme humain... Apologie de la définition bonaldienne de
     l'homme
1856: *Vicissitudes de l'anthropologie hippocratique*
1857: *Rappel des principes doctrinaux de la constitution de l'homme*
     énoncés par Hippocrate et démontrés par Barthez
1857: *Lettre de M. Lordat... au sujet d'un nouvel écorché*
1858: *Indispensable nécessité de la philosophie naturelle*
     expérimentale inductive

Bérard, Frédéric (1789–1829)

1818: *Essai sur les anomalies de la variole et de la varicelle*
1819: *Doctrine médicale de l'école de Montpellier* [1836]
1819: *Observations cliniques pour servir de preuves à la doctrine... de*
     Montpellier

1820: *Mémoire sur les avantages politiques et scientifiques du concours...*

1823: *Doctrine du rapport du physique et du moral*

1823: *Réponse de M. Bérard à M. Boisseau*

1826: *Discours sur les améliorations progressives de la santé publique*

1830: *Esprit des doctrines médicales de Montpellier*

Alquié, Alexis (1812–1864)

1845: *Examen du rapport relatif à la mort de la dame Saugeron d'Evreux*

1845: *Cours élémentaire de pathologie chirurgicale d'après la doctrine de Montpellier*

1846: *Précis de la doctrine médicale de l'école de Montpellier*

1850: *Des qualités et des devoirs du professeur de clinique chirurgicale*

1850: *Chirurgie conservatrice et moyens de restreindre l'utilité des opérations*

1850: *Traité élémentaire de pathologie médicale d'après la doctrine de Montpellier*

1852: *Clinique chirurgicale de l'Hôtel-Dieu de Montpellier*

1864: *Étude médicale et expérimentale de l'homicide par strangulation* [1865]

Bichat, Xavier (1771–1802)

1792: *De l'influence que la mort du poumon exerce sur la mort du coeur...*

1799: *Recherches physiologiques sur la vie et la mort* [1805, 1812, 1822, 1835, 1844, 1852, 1856, 1859, 1862, 1866, 1873]

1800: *Traité des membranes en général et des diverses membranes* [1816]

1801: *Traité d'anatomie descriptive* [1812, 1814, 1823, 1829]

1801: *Anatomie générale appliquée à la physiologie et à la médecine* [1812, 1818]

1825: *Anatomie pathologique*

Dugès, Antoine (1797–1838)

1823: *Essai physiologico-pathologique sur la fièvre... Suivi de l'histoire des maladies observées à l'hôpital des Enfants malades*

1826: *Manuel d'obstétrique, ou Traité de la science et de l'art des accouchements*

1833: *Traité pratique des maladies de l'utérus et de ses annexes...*

1838: *Traité de physiologie comparée de l'homme et des animaux*

Corvisart, Jean-Nicolas (1755–1821)

1797: *Aphorismes sur la connaissance et la curation des fièvres...*

1808: *Nouvelle méthode pour reconnaître les maladies de la poitrine...* [1838]

1814: **Journal de médecine, chirurgie, pharmacie*

1818: *Essai sur les maladies et les lésions organiques du coeur...* [1838]

Broussais, François (1772–1838)

1808: *Histoire des phlegmasies ou inflammations chroniques* [1816, 1816, 1838]

1819: *Leçons du Dr Broussais sur les phlegmasies gastriques...* [1823]

1821: *Examen des doctrines médicales et des systèmes de nosologie* [1834]

1822: **Annales de la médecine physiologique*

1823: *Traité de physiologie appliqué à la pathologie* [1834]

1824: *Le catéchisme de la médecine physiologique*

1826: *Discours préliminaire pour l'année 1826 des Annales de la médecine physiologique*

1826: *De la théorie médicale dite pathologique, ou Jugement de l'ouvrage de M. Prus*

1828: *Quelques mots sur les attaques... des Kanto-Platoniciens du Globe* [1828, 1829]

1829: *Développement des propositions relatives à la pathologie...* [1829]

1829: *Du prétendu éclectisme médical*

1831: *Cours de pathologie et de thérapeutique générales* [1834]

1832: *Le choléra-morbus épidémique* [1832, 1832]

1832: *Deux leçons du Prof. Broussais sur le choléra-morbus* [1832, 1832, 1832, 1832]

1832: *Méthode Broussais*

1832: *Lettre de M. Broussais sur le choléra-morbus*

1832: *Mémoire sur l'influence que les... médecins physiologistes ont exercée sur l'état de la médecine en France*

1832: *Mémoire sur la philosophie de la médecine*

1832: *De la meilleure manière de philosopher en médecine et des obstacles qui en retardent les progrès*

1836: *Cours de phrénologie*

1839: *De l'irritation et de la folie*

1840: *Recueil de mémoires de médecine, de chirurgie et de pharmacie militaire*

Laënnec, René (1781–1826)

1804: *Propositions sur la doctrine d'Hippocrate*

1805: *Mémoire sur les vers vésiculaires...*

1812: *Note sur une espèce de hernie... qu'on pourrait appeler extra-péritonéale*

1819: *De l'auscultation médiate, ou Traité du diagnostic des maladies du poumon et du coeur* [1826, 1828, 1837, 1879, 1893]

1826: *Notice des faits nouveaux obtenus par... l'auscultation médiate*

1884: *Traité inédit sur l'anatomie pathologique*

Magendie, François (1783–1855)

1809: *Examen de l'action de quelques végétaux sur la moelle épinière*

1813: *De l'influence de l'émétique sur l'homme et sur les animaux*

1813: *Mémoire sur l'usage de l'épiglotte dans la déglutition*

1813: *Mémoire sur le vomissement*

1816: *Précis élémentaire de physiologie* [1825, 1836]

1818: *Recherches physiologiques et médicales sur... la gravelle*

1819: *Recherches physiologiques et cliniques sur l'emploi de l'acide prussique*

1821: *Formulaire pour la préparation et l'emploi de plusieurs nouveaux médicaments* [1822, 1822, 1824, 1825, 1827, 1829, 1836]

1821: **Journal de physiologie expérimentale*

1822: *Recherches d'anatomie pathologique sur l'endurcissement du système nerveux*

1823: *Rapport à la Société royale des Sciences sur les planches anatomiques...*

1823: *Mémoire sur quelques découvertes récentes sur le... système nerveux* [1823]

1825: *L'ouïe et la parole...*

1825: *Anatomie des systèmes nerveux des animaux à vertèbres*

1828: *Mémoire physiologique sur le cerveau*

1832: *Leçons sur le choléra-morbus*

1832: *De l'or, de son emploi dans le traitement de la siphillis...*

1837: *Leçons sur les phénomènes physiques de la vie*

1838: *Leçons sur le sang et les altérations de ce liquide dans les maladies graves*

1839: *Leçons sur les fonctions et les maladies du système nerveux*

1841: *Rapport fait à l'Académie des Sciences...*

1842: *Recherches physiologiques et cliniques sur le liquide céphalorachidien*

1842: *Rapport... relatif à l'insufflation du poumon*

1845: *Rapport sur un bras artificiel* [1846, 1856]

1852: *Leçons faites au Collège de France pendant le semestre d'hiver 1851-1852*

1874: *Filtrage des eaux de Paris...* [1878]

Bernard, Claude (1813–1878)

1843: *Essai sur le suc gastrique et sur son rôle dans la nutrition*

1843: *Recherches anatomiques et physiologiques sur la corde du tympan*

1845: *Sur les phénomènes chimiques de la digestion*

1850: *Notice sur les travaux d'anatomie et de physiologie*

1852: *Notice sur les travaux de M. Claude Bernard* [1854]

1853: *Recherches sur une nouvelle fonction du foie considéré comme organe producteur de matière sucrée...* [1853]

1854: *Précis iconographique de médecine opératoire* [1870, 1873]

1854: *Recherches expérimentales sur le grand sympathique* [1878, 1890]

1855: *Leçons de physiologie expérimentale appliquée à la médecine* [1856, 1856]

1856: *Leçon d'ouverture du cours de médecine du Collège de France*

1856: *Mémoire sur le pancréas et sur le rôle du suc pancréatique*

1857: *Leçons sur les effets des substances toxiques et médicamenteuses* [1883]

1858: *Leçons sur la physiologie et la pathologie du système nerveux* [1858]

1858: *De la méthode expérimentale, de l'expérimentation et de ses perfectionnements*

1859: *Leçons sur les propriétés physiologiques et les altérations pathologiques des liquides de l'organisme* [1859]

1865: *Introduction à l'étude de la médecine expérimentale* [1898, 1898]

1867: *Rapports sur les progrès de la physiologie générale en France*

1872: *Leçons de pathologie expérimentale* [1880]

1872: *De la physiologie générale* [1872]

1875: *Leçons sur les anesthésiques et sur l'asphyxie*

1876: *Leçons sur la chaleur animale, sur les effets de la chaleur et de la fièvre*

1877: *Leçons sur le diabète et la glycogenèse animale*

1878: *Leçons sur les phénomènes de la vie communs aux animaux et aux végétaux* [1885]

1879: *Leçons de physiologie opératoire*

Bibliography

Ackerknecht, E.H. 1986. *La Médecine hospitalière à Paris (1794–1848)*. Paris: Éditions Payot.

Ackermann, R. 1985. *Data, Instruments, and Theory*. Princeton: Princeton University Press.

Akrich, M., Callon, M., Latour, B. 2006. *Sociologie de la traduction. Textes fondamentaux*. Paris: Presses de l'Ecole des Mines.

Albertus Magnus and the Sciences 1980. *Commemorative Essays*. Edited by J.A. Weisheipl, O.P. Toronto: Pontifical Institute of Medieval Studies.

Alquié, A. 1845. *Protestation en faveur de l'école de Montpellier*. Montpellier: J. Martel.

Alquié, A. 1846. *Précis de la doctrine médicale de l'école de Montpellier*. Montpellier: Ricard frères.

Andrews, F., ed. 1978. *Scientific Productivity: The Effectiveness of Research Groups in Six Countries*. London: Cambridge University Press.

Angelelli, I. 1970. "The techniques of disputation in the history of logic," *Journal of Philosophy* 67: 800–815.

Aristotle 2007. *Topiques*, Livres V–VIII, ed. et trad. J. Brunschwig. Paris: Les Belles Lettres.

———. 1992. *De Animalibus. Michael Scot's Arabic-Latin Translation. Part Three: Books xv–xix. Generation of Animals*, edited by A.M.I. van Oppenraaij. Leiden: E.J. Brill.

Aron, R. 1967. *Les Étapes de la pensée sociologique*. Paris: Gallimard.

Arnau Soler, E., Ayala, S., Sturm, T. 2013. "Cognitive externalism meets bounded rationality." *The Reach of Radical Embodied or Enactive Cognition*. Antwerp: University of Antwerp, June 17–19 2013.

Anonymous. 1478. *Larte de labbaccho,* copia anastatica a cura di G. Romano. Treviso: Longo e Zoppelli, 1969.

ATLAS 2012. "Observation of a new particle in the search for the standard model Higgs boson with the ATLAS Detector at the LHC." *Physics Letters B* 716, 1–29.

Augustine 2002. *S. Aureli Augustini opera (Corpus scriptorum ecclesiasticorum latinorum*, 12–74). Vindobonae, apud C. Geroldi Filium, 1887. *On Genesis*, translated by J.E. Rotelle, New York: Augustinian Heritage Institute.

Bachelard, G. 1938. *La Formation de l'esprit scientifique*. Paris: Librairie philosophique J. Vrin, 1972. English translation: *The Formation of the Scientific Mind*. San Francisco: Clinamen Press.

Bacon, R. 1964. *The 'Opus Maius' of Roger Bacon*, edited with Introduction and Analytical Table by J.H. Bridges. Frankfurt am Main: Minerva GmbH.

——. 1859. *Opera quaedam hactenus inedita*, vol. 1. Edited by J.S. Brewer. London: Longman, Green and Roberts.

Barlotta, P., and Dascal, M., eds. 2005. *Controversies and Subjectivity*. Amsterdam: John Benjamins Publishing Co.

Barnes, B. 1983. "On the conventional character of knowledge and cognition." Knorr-Cetina, K.D. and Mulkay, M., eds. *Science Observed: Perspectives on the Social Study of Science*. London: Sage, 19–51.

——. 1977. *Interests and Growth of knowledge*. London: Routledge and Kegan Paul.

——. and Bloor, D. 1981. "Relativism, rationalism and the sociology of knowledge." Hollis, M. and Lukes, S., eds. *Rationality and Relativism*. Oxford: Blackwell, 21–47.

Barthez, P.J. 1778. *Nouveaux élémens de la science de l'homme*. Montpellier: J. Martel.

Basil of Caesarea 1857. *S.P.N. Basilii Caesareae Cappadocia archiepiscopi, Opera omnia quae exstant... tomus primus*. Excudebatur et venit apud J.-P. Migne editorem... Paris.

Bayle, P. 1683. *Pensées diverses écrites à un Docteur de Sorbonne, à l'occasion de la Comete qui parut au mois de Décembre 1680*, Rotterdam, Reinier Leers. Critical edition by A. Prat. Paris: Société Nouvelle de Librairie et d'Édition, 1911.

Bechler, R.G. 2012. "Hansen versus Neisser: controvérsias científicas na 'descoberta' do bacilo da lepra," *História Ciências Saúde–Manguinhos* 19(3): 815–842.

Bechler, Z. 1974. "Newton's 1672 optical controversies: A study in the grammar of scientific dissent." Elkana, Y, ed. *The Interaction between Science and Philosophy*. Atlantic Highlands: Humanities Press, 115–142.

Ben-David, J. 1991. *Scientific Growth*, edited by G. Freudenthal. Berkeley, Los Angeles, Oxford: University of California Press. French translation: *Éléments d'une sociologie historique des sciences*, edited by G. Freudenthal. Paris: Presses universitaires de France, 1997.

Bérard F., 1819a. *Doctrine médicale de l'école de Montpellier et comparaison de ses principes avec ceux des autres écoles d'Europe*, Montpellier: Imprimerie J. Martel.

——. 1819b. *Observations cliniques pour servir de preuve à la doctrine médicale de l'école de Montpellier*. Montpellier: Imprimerie J. Martel.

Berger, P.L., and Luckmann, T. 1966. *Social Construction of Reality. A Treatise in the Sociology of Knowledge*. Garden City, NY: Anchor Books. French translation: *La Construction sociale de la réalité*. Paris: Méridiens Klincksieck, 1986.

Berjak, R., and Iqbāl, M. 2003. "Ibn Sīnā–al-Bīrūnī Correspondence, al-As'ilah wa'l-Ajwibah," *Islam and Science* 1: 91–98, 253–260.

Berthelot, J.-M. 2008. *L'Emprise du vrai*, Paris: PUF.

Besel, R.D. 2011. "Opening the 'black box' of climate change science: Actor-Network Theory and rhetorical practice in scientific controversies," *Southern Communication Journal* 76(2): 120–136.

Bloor, D. 1997. "Remember the 'Strong Programme'?" *Enquête* 5: 55–68.

———. 1992. "Left and right Wittgensteinians." Pickering, A., ed. *Science as Practice and Culture*. Chicago: Chicago University Press, 266–282.

———. 1983. *Wittgenstein: A Social Theory of Knowledge*. London: Macmillan.

———. 1976. *Knowledge and Social Imagery*, London: Routledge. New edition: London, The University of Chicago Press, 1991. French translation: *Sociologie de la logique, ou les limites de l'épistémologie*. Paris: Pandore, 1983.

———. 1973. "Wittgenstein and Mannheim on the sociology of mathematics." *Studies in the History and Philosophy of Science* 4(2): 173–191.

Boghossian, P. 2006. *The Fear of Knowledge. Against Relativism and Constructivism*. Oxford: Oxford University Press.

Boudon, R. 1995. *Le Juste et le Vrai. Études sur l'objectivité des valeurs et de la connaissance*. Paris: Fayard.

———. 1994. "Relativiser le relativisme: quand la sociologie réfute la sociologie des sciences." *Revue Tocqueville* 15(2), 109–131.

———. 1990. *L'Art de se persuader des idées douteuses, fragiles ou fausses*. Paris: Fayard.

———. 1986. *L'Idéologie*. Paris: Fayard.

———. and Clavelin, M. eds. 1994. *Le Relativisme est-il résistible? Regards sur la sociologie des sciences*. Paris: Presses universitaires de France.

Bourdieu, P. 1994. *Raisons pratiques. Sur la théorie de l'action*. Paris: Éditions du Seuil.

Bouveresse J. 1999. *Prodiges et vertiges de l'analogie. De l'abus des belles-lettres dans la pensée*. Paris: Raisons d'agir.

———. 1984. *Rationalité et cynisme*. Paris: Éditions de Minuit.

Boyer, A. 2000. "Philosophie des sciences." P. Engel, ed. *Précis de philosophie analytique*. Paris: Presses universitaires de France, 157–188.

———. 1978. *K.R. Popper: une épistémologie laïque?* Paris: Presses de l'École normale supérieure.

Brayrooke, D., and Lindblom, C. 1963. *A Strategy of Decision*. London: Collier-Macmillan.

Brenner, A. 1990. *Duhem. Science, réalité et apparence. La relation entre philosophie et histoire dans l'oeuvre de Pierre Duhem*. Paris: Librairie philosophique J. Vrin.

Brock, W.H. 2010. "The atomic debate revisited," *ACS Symposium Series* 1044: 59–64.

Brockliss, L.W. 1992. "La querelle entre les facultés de médecine de Montpellier et de Paris au XVIIe siècle." *7e Centenaire des universités de l'académie de Montpellier (1289–1989)*. Montpellier: Université de Montpellier I, 44–49.

Brossard, D. 2009. "Media, scientific journals and science communication: examining the construction of scientific controversies." *Public Understanding of Science* 18(3): 258–274.

Broussais F. 1829. *Examen des doctrines médicales et des systèmes de nosologie*, t. 3. Paris: Delaunay.

Bunge, M. 2012. *Evaluating Philosophies*, New York, Springer Verlag.

———. 2006. *Chasing reality. Strive over realism*. Toronto: University of Toronto Press.

———. 1999. *The Sociology-Philosophy connection*, New Brunswick: Transaction Publishers.

———. 1992. "A critical examination of the new sociology of science, part II." *Philosophy of the Social Sciences* 22: 46–76.

———. 1991. "A critical examination of the new sociology of science, part I." *Philosophy of the Social Sciences* 21: 524–560.

Burnham, F.B. 1974. "The More-Vaughan controversy: The revolt against philosophical enthusiasm," *Journal of the History of Ideas* 35: 33–49.

Caillé, A. 1989. *Critique de la raison utilitaire*. Paris: Éditions La Découverte.

Callebaut, A. 1925. "Jean Pecham, OFM et l'augustinisme: aperçus historiques (1263–1285)." *Archivum franciscanum historicum* 18: 441–472.

Callon, M. 1981. "Pour une sociologie des controverses technologiques." *Fundamenta Scientiae* 2: 381–399.

———., ed. 1989. *La Science et ses réseaux. Genèse et circulation des faits scientifiques*. Paris: Éditions La Découverte.

———. and Latour, B., eds. 1991. *La Science telle qu'elle se fait*. Paris: Éditions La Découverte.

———. and Latour, B., eds. 1985. *Les Scientifiques et leurs alliés*. Paris: Pandore.

Canguilhem, G. 1977. *Idéologie et rationalité dans les sciences de la vie*. Paris: Librairie philosophique J. Vrin.

———. 1968. *Études d'histoire et de philosophie des sciences*. Paris: Librairie philosophique J. Vrin.

Cantor, G.N. and Hodge, M.J.S., eds. 1981. *Conceptions of Ether: Studies in the History of Ether Theories, 1740–1900*. Cambridge: Cambridge University Press.

Cantor-Coquidé, M. 1992. *Félix-Archimède Pouchet, savant et vulgarisateur*. Thèse de doctorat. Orsay: Université Paris-Sud.

Caplow, T. 1968. *Two Against One. Coalitions in Triads*. New York: Prentice Hall, French translation: *Deux contre un. Les coalitions dans les triades*. Paris: Armand Colin, 1968.

Carnap, R. 1936. "Testability and meaning". *Philosophy of Science* 3: 419–471, 4: 2:40.

Carroll, L. 1895. "What the Tortoise said to Achilles." *Mind* 4: 278–280.

Cartwright, N. 1983. *How the Laws of Physics Lies*. Oxford: Clarendon Press.

Chateauraynaud, F. 1991. "Forces et faiblesses de la nouvelle anthropologie des sciences." *Critique* 529/530: 465–466.

Chrestien, A. 1860. *Lettre à son Exc. M. Rouland, Ministre de l'Instruction publique et des Cultes, sur l'antagonisme qui a toujours existé entre la faculté de médecine de Montpellier et celle de Paris.* Cette: Typographie G. Bonnet.

Chrestien, A. 1830. "À M. le Rédacteur du Mémorial, etc." *Mémorial des Hôpitaux du Midi et de la Clinique de Montpellier* 2: 57–59.

Clark, A. 2000. "Cognitive incrementalism: The big issue." *Behavioral and Brain Sciences* 23(4): 536–537.

Clavelin, M. 1968. *La Philosophie naturelle de Galilée. Essai sur les origines et la fonction de la mécanique classique.* Paris: A. Colin.

Cohen, I.B. 1939. "Roemer and the first determination of the velocity of light (1676)." *Isis* 31: 327–379.

Cole, S. 1996. "Voodoo Sociology: recent developments in the sociology of science." Gross, P.K., Levitt, N. and Lewis, M.W., eds. *The Flight from Science and Reason.* New York: New York Academy of Sciences, 274–287.

Collingridge, D. 1989. "Incremental decision making in technological innovations: what role for science?" *Science, Technology and Human Values* 14(2): 141–162.

———. and Reeve, C. 1986. "Science and policy." *Bulletin of Science, Technology and Society* 6: 356–372.

Collins, H.M. (1981). "Stages in the Empirical Programme of Relativism." *Social Studies of Science* 11(1): 3–10.

———. (1974). "The TEA set: Tacit knowledge and scientific networks." *Science Studies* 4: 165–186.

———. and Pinch, T.J. (1993). *The Golem. What Everyone Should Know about Science,* Cambridge: Cambridge University Press. French translation: *Tout ce que vous devriez savoir sur la science.* Paris: Éditions du Seuil, 1994.

———. and Pinch, T.J. (1991). "En parapsychologie, rien ne se passe qui ne soit scientifique..." Callon, M. and Latour, B., eds. *La Science telle qu'elle se fait.* Paris: Éditions La Découverte, 297–343 [1979].

———. and Cox, G. (1976). "Recovering relativity: Did prophecy fail?" *Social Studies of Science* 6: 423–445.

Condorcet, J.A. Caritat de 1974. *Mathématique et société.* Choix de textes et commentaire par Roshdi Rashed. Paris: Hermann.

———. 1788. *Essai sur la constitution et les fonctions des assemblées provinciales, Où l'on trouve un Plan pour la Constitution & l'Administration de la France, Tome Premier.* s.l., s.n.

Coser, L.A. 1956. *The Functions of Social Conflict,* New York, The Free Press. French translation: *Les Fonctions du conflit social.* Paris: Presses universitaires de France, 1982.

Crellin, J.K. 1981. "Pouchet." C.C. Gillispie, ed. *Dictionary of Scientific Biography.* New York: Charles Scribner's Sons, 11: 109–110.

———. 1966. "Airborne particles and the germ theory: 1860–1880." *Annals of Science* 22: 49–60.

Crombie, A.C. 1994. *Styles of Scientific Thinking in the European Tradition,* 3 vols. London: Duckworth and Co.

——. 1953. *Robert Grosseteste and the Origins of Experimental Science (1100–1700).* Oxford: Clarendon Press.

——. 1952. *Augustine to Galileo. The History of Science A.D. 400–1650.* London: Falcon. French translation: *Histoire des sciences de Saint Augustin à Galilée (400–1650).* Paris: Presses universitaires de France, 1959.

Crosland, M. 1992. *Science under Control. The French Academy of Science, 1795–1914.* Cambridge: Cambridge University Press.

Crowe, M.J. 1990. "Duhem and history and philosophy of mathematics." *Synthese* 83: 431–437.

Dagognet, F. 1967. *Méthodes et doctrines dans l'oeuvre de Pasteur.* Paris: Presses universitaires de France.

Darling, K.M. 2003. "Motivational realism: The natural classification for Pierre Duhem," *Philosophy of Science* 70(5): 1125–1136.

Darmon, P. 1995. *Pasteur.* Paris: Fayard.

Dascal, M. 2000. "Controversies and epistemology." Tian Yu Cao, ed., *Philosophy of Science.* Philadelphia: Philosophers Index Inc., 159–192. English translation of "Epistemología, controversias y pragmática." *Isegoría* 12: 8–43.

——. 1998. "The study of controversies and the theory and history of science." *Science in Context* 11(2): 147–154.

——., Racionero, Q. and Cardoso, A. 2006. *Leibniz, Gottfried Wilhelm: The Art of Controversies.* Dordrecht: Springer.

Debarnot, M.-T. 1985. *Kitāb maqālīd 'ilm al-hay'a. La trigonométrie sphérique chez les arabes de l'est à la fin du Xe siècle.* Damas: Institut français de Damas.

De Freitas, R.S. and Pietrobon, R. 2010. "Why care about scientific controversies? Continuities and discontinuities in the history of science." *Journal of Historical Sociology* 23(4): 501–516.

Delpech J. 1830. "Variétés." *Mémorial des Hôpitaux du Midi et de la Clinique de Montpellier* 2: 318–319.

Demeter, T. and Zemplén, G.A. 2010. "Being charitable to scientific controversies: on the demonstrativity of Newton's experimentum crucis." *The Monist* 93(4): 640–656.

Desanti, J.-T. 1975. *La Philosophie silencieuse, ou critique des philosophies de la science.* Paris: Éditions du Seuil.

Descartes, R. 1826. *Règles pour la direction de l'esprit,* in *Oeuvres de Descartes,* vol. 11, publiées par V. Cousin. Paris: F.-G. Levrault, 1826, d'après les *Regulae ad directionum ingenii,* in *Opuscula posthuma, physica et mathematica,* Amsterdam, Jansson, 1701.

De Young, G. 2001. "The Ashkāl al-Ta'sīs of al-Samarqandī: A Translation and Study." *Zeitschrift für Geschichte der arabisch-islamischen Wissenschaften* 14: 57–117.

Dilgan, H. 1975. "Samarqandī, Shams al-Dīn Muhammad Ibn Ashraf al-Husaynī al-." C.C. Gillispie, ed. *Dictionary of Scientific Biography.* New York: Scribner, vol. 12, 19.

Dilgan, Ḥ. 1960. "Démonstration du Vᵉ postulat d'Euclide par Shams ed-Dīn Samarqandī." *Revue d'histoire des sciences* 13: 191–196.

Djebbar, A. 2001. *Une Histoire de la science arabe*. Paris: Éditions du Seuil.

Doesch, R.N. 1962. "Early American experiments on 'spontaneous generation' by Jeffries Wyman (1814–1874)." *Journal of the History of Medicine and Allied Sciences* 17: 326–332.

Downing, K.L. 2010. "A brief introduction to situated and embodied artificial intelligence." http://www.idi.ntnu.no/emner/it3708/lectures/notes/intro.pdf

Dubois, M. 1998. "L'affaire Sokal et la question de la sociologie relativiste des sciences." *Revue française de Sociologie* 39(2): 391–418.

Duhem, P. 1913–1959. *Le Système du monde. Histoire des doctrines cosmologiques de Platon à Copernic* (tomes 6, 7, 8). Paris: Hermann.

———. 1908. *Sôzein ta phainomena. Essai sur la notion de théorie physique, de Platon à Galilée*. Paris: Librairie philosophique J. Vrin, 1980. English translation: *To Save the Phenomena. An Essay on the Idea of Physical Theory*. Chicago: University of Chicago Press, 1969.

———. 1905. *La Théorie physique: son objet, sa structure*. Paris: Librairie philosophique J. Vrin, 1985. English translation: *The Aim and Structure of Physical Theory*. Princeton: Princeton University Press, 1991.

———. 1894a. "Quelques réflexions au sujet de la physique expérimentale." *Revue des Questions scientifiques* 36: 179–229.

———. 1894b. "Les théories de l'optique." *Revue des Deux Mondes* 123: 94–125.

Dulieu, L., ed. 1990. *La Médecine à Montpellier du XIIᵉ au XXᵉ siècle* (30ᵉ Congrès international d'histoire de la médecine, septembre 1986). Paris: Hervas.

———. 1983–1990. *La Médecine à Montpellier, III: L'âge classique*, 1: 1983, 2: 1986. IV: *De la Première à la Troisième République*, 1: 1988, 2: 1990. Avignon: Les Presses universelles.

Dummett, M. 1959. "Wittgenstein's philosophy of mathematics." *Philosophical Review* 68: 324–348.

Dupuy, J.-P. and Livet, P., eds. 1997. *Les Limites de la rationalité*, tome 1: *Rationalité, éthique, cognition*. Paris: Éditions La Découverte.

Durkheim, E. 1925. *Sociologie et philosophie*. Paris: Presses universitaires de France, 1963. English translation: *Sociology and Philosophy*. London: Routledge, 2009.

———. 1912. *Les Formes élémentaires de la vie religieuse*. Paris: Presses universitaires de France, 1960. English translation: *The Elementary Forms of the Religious Life*. Oxford: Oxford University Press, 2001.

———. 1895. *Les Règles de la méthode sociologique*. Paris: Presses universitaires de France, 1937. English translation: *The Rules of Sociological Method*. New York: The Free Press, 1982.

———. and Mauss, M. 1902. "De quelques formes primitives de classification." *L'Année sociologique* 6: 1–72.

Eastwood, B.S. 1989. *Astronomy and Optics from Pliny to Descartes: Texts, Diagrams and Conceptual Structures*. London: Variorum Reprints.

Eberlein, G.L. 1994. "La 'nouvelle sociologie des science': un nouvel irrationalisme?" Boudon, R. and Clavelin, M., eds. *Le Relativisme est-il résistible?* Paris: Presses universitaires de France, 131–143.

Engel, P. ed. 2000. *Précis de philosophie analytique*. Paris: Presses universitaires de France.

——. 1998. *La Vérité. Réflexions sur quelques truismes*. Paris: Hatier.

——. 1989. *La Norme du vrai*. Paris: Gallimard.

Engelhardt, H.T. and Caplan, A.L. eds. 1987. *Scientific Controversies: Case Studies in the Resolution of Closure of Disputes in Science and Technology*. Cambridge: Cambridge University Press.

Fabiani, J.-L. 1997. "Controverses scientifiques, controverses philosophiques. Figures, positions, trajets." *Enquête* 5: 11–34.

Farley, J. 1972. "The spontaneous generation controversy (1700–1860): The origin of parasitic worms." *Journal of the History of Biology* 5: 95–125.

——. and Geison, G. 1974. "Science, Politics, and Spontaneous Generation in Nineteenth-Century France: The Pasteur-Pouchet Debate." *Bulletin of the History of Medicine* 48: 161–198. French translation: "Le débat entre Pasteur et Pouchet: science, politique et génération spontanée au XIXe siècle en France." Callon, M. and Latour, B., eds., *La Science telle qu'elle se fait*. Paris: Éditions La Découverte, 1991, 87–145.

Farrall, L.A. 1975. "Controversy and conflict in science: A case study—the English Biometric School and Mendel's laws." *Social Studies of Science* 5: 269–301.

Fazlıoğlu, İ. 2007. "Samarqandī: Shams al-Dīn Muḥammad ibn Ashraf al-Ḥusaynī al-Samarqandī." T. Hockey et al., eds. *The Biographical Encyclopedia of Astronomers*. New York: Springer, 1008.

Felipe, D. 1991. *The Post-Medieval Ars Disputandi*, PhD Dissertation. Austin: University of Texas.

Festinger, L. 1957. *A Theory of Cognitive Dissonance*. Evanston: Row and Peterson.

Feyerabend, P. 1975. *Against Method. Outline of an Anarchist Theory of Knowledge*. London: New Left Books. French translation: *Contre la méthode. Essai d'une théorique anarchiste de la connaissance*. Paris: Éditions du Seuil, 1979.

Figatelli, G.M. 1678. *Trattato aritmetico*. Venice: Stefano Curti.

Flap, H. 1997. "The conflicting loyalties theory." *L'Année sociologique* 47(1): 183–215.

Fleck, L. 1935. *Genesis and Development of a Scientific Fact*. Chicago: Chicago University Press.

Fourez, G. 1996. *La Construction des sciences. Introduction à la philosophie et à l'éthique des sciences*. Bruxelles: De Boeck Université.

Frankel, E. 1976. "Corpuscular optics and the wave theory of light. The science and politics of a revolution." *Social Studies of Science* 6: 141–184.

Frankel, H. 1987. "The continental drift debate." Engelhardt, H.T. and Caplan, A.L., eds. *Scientific Controversies. Case Studies in the Resolution and Closure of Disputes in Science and Technology*. New York: Cambridge University Press, 203–248.

Freudenthal, G. 1998. "Controversy." *Science in Context* 11(2): 155–160.

———. 1984. "The role of shared knowledge in science: the failure of constructivist programme in the sociology of science." *Social Studies of Science* 14: 285–295.

Galileo, G. 1970. *Discours et démonstrations concernant deux sciences nouvelles.* Introduction, traduction et notes de Maurice Clavelin. Paris: Colin [1638].

Galison, P. 1987. *How Experiments Ends.* Chicago: University of Chicago Press.

Gálvez, A. 1988. "The role of the French Academy of Sciences in the clarification of the issue of spontaneous generation in the mid-nineteenth century." *Annals of Science* 45: 345–365.

Gärdenfors, P., ed. 1992. *Belief Revision.* Cambridge: Cambridge University Press.

Garrison, F.H. 1929. *An Introduction to the History of Medicine.* Philadelphia/ London: Saunders.

Gazette 1846. "Génie et perpétuité de l'école de Montpellier." *Gazette médicale de Paris* 32: 613–614.

———. 1834. "Concours de Montpellier. Incidens divers. Protestations de la majorité des concurrens." *Gazette médicale de Paris* 2: 116–118.

Geison, G. 1981. "Pasteur." C.C. Gillispie, ed., *Dictionary of Scientific Biography.* New York: Charles Scribner's Sons, 9: 350–414.

Giere, R.N. 1988. *Explaining Science. A Cognitive Approach.* Chicago: The University of Chicago Press.

Gieryn, T. 1982. "Relativist/constructivist programmes in the sociology of science: Redundance and retreat." *Social Studies of Science* 12: 279–297.

Gillispie, C.C., ed. 1981. *Dictionary of Scientific Biography*, 8 vols. New York: Charles Scribner's Sons.

Gingras, Y. 2013. *Sociologie des sciences*, Paris, PUF.

Gochet, P. 1978. *Quine en perspective. Essai de philosophie comparée.* Paris: Flammarion.

Goodwin, W. 2013. "*Structure* and scientific controversies." *Topoi* 32: 101–110.

Goubert, J.-P. and Rey R. eds. 1993. *Atlas de la Révolution française*, tome 7: *Médecine et santé.* Paris: Éditions de l'EHESS.

Granger, G.G. 1969. *Wittgenstein.* Paris: Seghers.

———. 1956. *La Mathématique sociale du Marquis de Condorcet.* Paris: Presses universitaires de France.

Gross, P.K., Levitt, N., and Lewis, M.W., eds. 1996. *The Flight from Science and Reason.* New York: New York Academy of Sciences.

Grosseteste, R. 1912. *De Luce seu de inchoatione formarum, De lineis angulis et figuris seu de fractionibus et reflexionibus radiorum, De iride seu de iride et speculo.* L. Baur, Die philosophischen Werke des Robert Grosseteste, *Beiträge zur Geschichte der Philosophie des Mittelalters*, 9: 1–778.

Hacking, I. 1992a. "Statistical language, statistical truth, and statistical reason." McMullin, E. ed., *The Social Dimensions of Science.* Notre Dame: University of Notre Dame Press, 130–157.

————. 1992b. "The self-vindication of the laboratory sciences." Pickering, A. ed., *Science as Practice and Culture*. Chicago: Chicago University Press, 29–64.

Hahn, R. 1986. *The Anatomy of a Scientific Institution: The Paris Academy of Sciences, 1666–1803*. French translation: *Anatomie d'une institution. L'Académie des Sciences de Paris, 1666–1803*. Paris: Éditions des Archives contemporaines, 1994.

Hamerla, R. 2003. "Edward Williams Morley and the atomic weight of oxygen; the death of Prout's hypothesis revisited." *Annals of Science* 60(4): 351–372.

Harwood, J. 1976–1977. "The race-intelligence controversy: A sociological approach, 1. Professional factors, 2. External factors." *Social Studies of Science* 6: 369–394, 7: 1–30.

Helvetius, C.A. 1775. *Les Progrès de la Raison dans la recherche du Vrai*, Londres, s.n.

Herder, J.G. von 1784–1791. *Ideen zur Philosophie der Geschichte der Menschheit*, 4 vols., Leipzig: Hartknoch. French translation: *Idées sur la philosophie de l'histoire de l'humanité*, 4 vols. Paris: Berger Levrault, 1827.

————. 1774. *Auch eine Philosophie der Geschichte zur Bildung der Menschheit*. Riga, s.n. French translation: *Une autre philosophie de l'histoire*. Paris: Aubier-Montaigne, 1964.

Hetherington, N.S. 1993. *Cosmology. Historical, Litterary, Philosophical, Religious and Scientific Perspectives*. New York: Garland Publishers, Inc.

Hogendijk, J.P. 1984. "Greek and Arabic Constructions of the Regular Heptagon." *Archive for History of Exact Sciences* 30: 197–330.

Hollis, M. 1992. "Social thought and social action." McMullin, E., ed., *The Social Dimensions of Science*. Notre Dame: University of Notre Dame Press, 68–84.

————. and Lukes, S., eds. 1981. *Rationality and Relativism*. Oxford: Blackwell.

Holton, G. 1993. *Science and Anti-Science*. Cambridge: Harvard University Press. French translation: *Science en gloire, science en procès*. Paris: Gallimard, 1998.

————. 1982. *L'Invention scientifique*. Paris: Presses universitaires de France.

Hookway, C. 1992. *Quine*. Bruxelles: De Boeck Université.

Horwitz Westbrook, R. 1981. "Needham." Gillispie, C.C., ed., *Dictionary of Scientific Biography*. New York: Charles Scribner's Sons, 10: 9–10.

Hoskin, M.A. 1976. "The 'Great Debate': what really happened." *Journal for the History of Astronomy* 7: 169–182.

Ifrah, G. 1981. *Histoire universelle des chiffres*, 2 vols. Paris: Robert Laffont.

IPCC 2001. *Summary for Policymakers. Climate Change 2001: Impacts, Adaptation, and Vulnerability*. A report of Working Group II of the IPCC.

————. 2013. *Summary for Policymakers. Climate Change 2013: The Physical Science Basis*. Contribution of Working Group I to the Fifth Assessment Report of the IPCC.

Isambert, F.-A. 1994. "Après l'échec du 'programme fort', une sociologie du contenu de la science reste-t-elle possible?" Boudon, R. and Clavelin, M. eds., *Le Relativisme est-il résistible? Regards sur la sociologie des sciences*. Paris: Presses universitaires de France, 51–76.

———. 1985. "Un 'programme fort' en sociologie de la science?" *Revue française de sociologie* 26: 481–508.

Jensen, E. 2012. "Scientific controversies and the struggle for symbolic power." B. Wagoner, E. Jensen and J.A. Oldmeadow, eds., *Culture and Social Change: Transforming Society Through the Power of Ideas*. Charlotte, NC: IAP-Information Age Publishing Inc., 161–183.

Julia, D. 1992. "L'université de médecine de Montpellier à l'époque moderne." *7e Centenaire des universités de l'académie de Montpellier (1289–1989)*. Montpellier: Université de Montpellier I, 26–38.

———. and Revel, J. 1989. *Les Universités européennes du XVIe au XVIIIe siècle, Histoire sociale des populations étudiantes*, tome 2. Paris: Éditions de l'EHESS.

Junges, A.L. 2013. "Desacordo racional e controvérsia científica." *Scientiae Studia* 11(3): 613–635.

Karabela, M.K. 2010) *The Development of Dialectic and Argumentation Theory in Post-Classical Islamic Intellectual History*, PhD Dissertation. Montreal: McGill University.

Kim, K.M. 1994. *Explaining Scientific Consensus: the Case of Mendelian Genetics*. New York: Guilford Publications.

Kitcher, P. 1992. "Authority, deference, and the role of individual reason." McMullin, E. ed., *The Social Dimensions of Science*. Notre Dame: University of Notre Dame Press, 244–271.

Knorr Cetina K. 1995. "Laboratory Studies: The Cultural Approach to the Study of Science." Jasanoff S. et al., eds., *Handbook of Science and Technology Studies*. London: Sage, 140–166.

———. 1981. *The Manufacture of Knowledge. An Essay in the Constructivist and Contextual Nature of Science*, Oxford: Pergamond Press.

———. 1977. "Producing and reproducing knowledge: Description or construction?" *Social Science Information* 16: 101–126.

———., Krohn, R., and Whitley, R., eds. 1981. *The Social Process of Scientific Investigation*. Dordrecht: Reidel.

Koertge, N., ed. 1998. *A House Built on Sand: Exposing Postmodernist Myths about Science*. New York: Oxford University Press.

Koyré, A. 1956. "L'hypothèse et l'expérience chez Newton." *Bulletin de la Société française de Philosophie* 50: 60–97.

Kuhn, T. 1980. "La tension essentielle: tradition et innovation dans la recherche scientifique." P. Jacob, ed., *De Vienne à Cambridge*. Paris: Gallimard, 303–320 [1959].

———. 1962. *The Structure of Scientific Revolutions*. Chicago: University of Chicago Press. French translation: *La Structure des révolutions scientifiques*. Paris: Flammarion, 1972.

Kühnholtz, H. 1843. *Paris et Montpellier sous le rapport de la philosophie médicale ...* Paris: J.-B. Baillière et al. / Montpellier: L. Castel Libraire-Editeur.

Kurrer, K.E. 2008. "Twelve scientific controversies in mechanics and theory of structures." K.E. Kurrer, *The History of the Theory of Structures: From*

Arch Analysis to Computational Mechanics, Berlin: Ernst & Sohn Verlag für Architektur und technische Wissenschaften GmbH, 674–690.

Laënnec, R.T.H. 1804. *Propositions sur la doctrine d'Hippocrate, relativement à la médecine-pratique*. Paris: Didot.

Lakatos, I. 1980. *The Methodology of Scientific Research Programmes*, edited by J. Worrall and G. Currie. Cambridge: Cambridge University Press. French translation: *Histoire et méthodologie des sciences. Programmes de recherche et reconstruction rationnelle*. Paris: Presses universitaires de France, 1994.

———. 1970. "Falsification and the Methodology of Scientific Research Programmes." I. Lakatos a,d A.E. Musgrave, eds., *Criticism and the Growth of Knowledge*, Cambridge: Cambridge University Press, 91–196.

Latour, B. 1992. "One more turn after the social turn." McMullin, E. ed., *The Social Dimensions of Science*. Notre Dame: University of Notre Dame Press, 272–294.

———. 1989. "Pasteur et Pouchet: hétérogenèse de l'histoire des sciences." M. Serres, *Éléments d'histoire des sciences*. Paris: Bordas, 423–445.

———. 1988. "A relativistic account of Einstein's relativity." *Social Studies of Science* 18: 3–44.

———. 1987. *Science in Action. How to Follow Scientists and Engineers Through*. Cambridge: Harvard University Press. French translation, *La Science en action*. Paris: Éditions La Découverte, 1989.

———. 1984. *Les Microbes – Guerre et Paix*. Suivi de *Irréductions*. Paris: Métailié and Pandore.

———. and Callon, M., eds. 1991. *La Science telle qu'elle se fait*. Paris: La Découverte.

———. and Woolgar, S. 1979. *Laboratory Life: The Social Construction of Scientific Facts*, Bervely Hills: Sage. New edition: Princeton: Princeton University Press, 1986. French translation: *La Vie de laboratoire*. Paris: La Découverte, 1988.

Laudan, L. 1990. *Science and Relativism*. Chicago: University of Chicago Press.

———. 1981. "The pseudo-science of science?" *Philosophy of the Social Sciences* 11: 173–198.

Lavabre-Bertrand, T. 1992a. *La Philosophie médicale de l'École de Montpellier au XIXe siècle*, thèse de doctorat. Paris, EPHE IVe section.

———. 1992b. "La philosophie médicale de l'École de Montpellier au XVIIIe et XIXe siècles." *7e Centenaire des universités de l'académie de Montpellier (1289–1989)*. Montpellier: Université de Montpellier I, 63–69.

Law, J. 1989. "Le laboratoire et ses réseaux." M. Callon, ed., *La Science et ses réseaux. Genèse et circulation des faits scientifiques*. Paris: La Découverte, 117–148.

Leclerc, G. 1996. *Histoire de l'autorité. L'assignation des énoncés culturels et la généalogie de la croyance*. Paris: Presses universitaires de France.

Lécuyer, B.-P. 1989. "La sociologie des sciences." *L'Univers philosophique*. Paris: Presses universitaires de France, 942–948.

———. et al. 1988. "Sociologie des sciences et des techniques." *L'Année Sociologique*, 38: 381–411.

Lemaine, G. and Matalon, B. 1985. *Hommes supérieurs, hommes inférieurs. La controverse sur l'hérédité de l'intelligence*. Paris: Armand Colin.

Leverington, A. 1995. *A History of Astronomy from 1890 to the Present*. London: Springer.

Lewis, D.K. 1969. *Convention. A Philosophical Study*. Cambridge: Harvard University Press.

Li, T. 2011. "The logical inference in scientific controversies. Einstein-Bohr controversies in quantum revolution." *Studies in Dialectics of Nature* 12: 34–39.

———. 2010. "The rhetorical model of solving scientific controversies." *Social Sciences in Ningxia* 4: 141–144.

Lindberg, D.C. 1996. *Roger Bacon and the Origins of Perspectiva in the Middle Ages*. A Critical Edition and English Translation, with Introduction and Notes. Oxford: Oxford University Press.

———. 1995. "Medieval science and its religious context." *Osiris* 10: 61–79.

———. 1983. *Roger Bacon's Philosophy of Nature: A Critical Edition with English Translation, Introduction and Notes of the 'De multiplicatione specierum' and 'De speculi comburentibus'*. Oxford: Clarendon Press.

———. 1978. "The intromission-extramission controversy in Islamic visual theory: Alkindi vs Avicenna." P.K. Machamer and R.G. Turnbull, eds., *Studies in Perception. Interrelations in the History of Philosophy and Science*. Colombus: Ohio State University Press, 137–159.

———. 1976. *Theories of Vision from Al-Kindi to Kepler*. Chicago and London: The University of Chicago Press.

———. 1971. "Lines of influence in the thirteenth century optics: Bacon, Witelo, and Pecham." *Speculum* 46: 66–83.

———. 1970. *John Pecham and the Science of Optics. Perspectiva Communis*. Edited with an Introduction, English Translation, and Critical Notes. Madison: The University of Wisconsin Press.

Lloyd, G.E. 2003. "Les concepts de vérité en Grèce et en Chine anciennes." J.-P. Changeux, ed., *La Vérité dans les sciences*. Paris: Odile Jacob, 49–60.

Lordat, J. 1852. *Accord de la Doctrine Anthropologique de Montpellier avec ce que demandent les Lois, la Morale publique et les Enseignements religieux prescrits par l'Etat*. Montpellier: J. Martel.

———. 1842. *Apologie de l'Ecole médicale de Montpellier, en réponse à la lettre écrite par M. Peisse à M. le Professeur Lordat*. Montpellier: L. Castel.

———. 1840. "Fragments de philosophie, par M. William Hamilton." *Journal de la Société de Médecine-pratique de Montpellier* 1: 391–421.

Lynch, M. 1993. *Scientific Practice and Ordinary Action. Ethnomethodology and Social Studies of Science*. Cambridge: Cambridge University Press.

———. 1992. "Extending Wittgenstein." Pickering, A. ed., *Science as Practice and Culture*. Chicago: Chicago University Press, 215–265.

———. 1985. *Art and Artifact in Laboratory Science*, London, Routledge and Kegan Paul.

Machamer, P.K., Pera, M. and Baltas, A., eds. 2000. *Scientific Controversies: Philosophical and Historical Perspectives*. Oxford/New York: Oxford University Press.

MacKenzie, D. 1993. "Negociating arithmetic, constructing proof: the sociology of mathematics and information technology." *Social Studies of Science* 23: 37–65.

———. 1978. "Statistical theory and social interests: a case study." *Social Studies of Science* 8(1): 35–83. French translation: "Comment faire une sociologie de la statistique." Callon, M. and Latour, B., eds., *La Science telle qu'elle se fait*. Paris: Éditions La Découverte, 1991, 200–261.

Macklin, R. 1987. "Forms and norms of closure." Engelhardt, H.T. and Caplan, A.L., eds., *Scientific Controversies*. New York: Cambridge University Press, 615–624.

Mannheim, K. 1952. *Essays on the Sociology of Knowledge*. London: Routledge and Kegan Paul.

———. 1936. *Ideology and Utopia*. London: Routledge. French translation: *Idéologie et Utopie*. Paris: Marcel Rivière, 1956.

al-Maqdisī 1899–1919. *Kitāb al-bad' wa-al-ta'rīkh*. French translation: *Le Livre de la création et de l'histoire*, ed. C. Huart, 5 vols. Paris: Ernest Leroux.

Marsh, N., Svensmark, H. 2000. "Cosmic rays, clouds and climate." *Space Science Review* 94: 215–230; preprint: www.dsri.dk/~hsv

Masoumi Hamedani, H. 2006. *L'Optique et la physique céleste: L'oeuvre optico-cosmologique d'Ibn al-Haytam*, 2 vols. Doctoral dissertation, Paris, Université Paris 7.

Matalon, B. 1986. "Sociologie des sciences et relativisme." *Revue de synthèse* 4e série, 3: 267–290.

McMullin, E., ed. 1992. *The Social Dimensions of Science*. Notre Dame: University of Notre Dame Press.

———. 1987. "Scientific controversy and its termination." Engelhardt, H.T. and Caplan, A.L., eds., *Scientific Controversies*. New York: Cambridge University Press, 49–91.

Medicus, F.C. 1774. *Von der Lebens Kraft*. Manheim: Akademische Buchdrucker.

Mendelsohn, E. 1987. "The political anatomy of controversy in the sciences." Engelhardt, H.T. and Caplan, A.L., eds., *Scientific Controversies*. New York: Cambridge University Press, 93–124.

Merton, R.K. 1973. *The Sociology of Science. Theoretical and Empirical Investigations*. Edited with an Introduction by Norman W. Storer. Chicago, London: The University of Chicago Press.

———. 1953. *Éléments de méthode sociologique*. Paris: Plon.

———. 1947. "La sociologie de la connaissance." Gurvitch, G. and Moore, W.E., eds., *La Sociologie au XXe siècle*. Paris: Presses universitaires de France, 366–405.

————. 1938. *Science, Technology and Society in Seventeenth-century England*. Bruges: St. Catherine Press Ltd.

————. 1937. "The sociology of knowledge." *Isis* 27: 493–503.

Meyerhof, M. 1928. *Kitāb al-'ashr maqālāt fī al-'ayn, al-mansūb li-Ḥunayn ibn Isḥāq / The Book of the Ten Treatises on the Eye Ascribed to Hunain Ibn Is-hâq*, edited from the only two known manuscripts, with an English translation and glossary, by Max Meyerhof. Cairo: Government Press.

Miller, L.B. 1984. *Islamic Disputation Theory. A Study of the Development of Dialectic in Islam from the Tenth through Fourteenth Centuries*, PhD Dissertation. Princeton University.

Moore, A. and Stilgoe, J. 2009. "Experts and anecdotes. The role of 'anecdotal evidence' in public scientific controversies." *Science, Technology & Human Values* 34(5): 654–677.

Montesquieu, C. de Secondat 1941. *Cahiers 1716–1755*. Paris, Grasset.

Moraux, P. 1968. "La joute dialectique d'après le huitième livre des Topiques." G.E.L. Owen, ed., *Aristotle on Dialectic*. Oxford: Clarendon Press, pp. 277–311.

Moussy, H. 1992. "L'école de santé de Montpellier, 1794–1803." *7e Centenaire des universités de l'académie de Montpellier (1289–1989)*. Montpellier: Université de Montpellier I, 59–62.

Mugler, C. 1964. *Dictionnaire historique de la terminologie optique des Grecs*. Paris: Librairie C. Klincksieck.

Mulkay, M. 1980. "Sociology of science in the West." *Current Sociology* 28(3): 1–184.

————. 1979. *Science and the Sociology of Knowledge*. London: Allen and Unwin.

Naudé, G. 1653. *Apologie pour tous les grands personnages qui ont esté faussement soupçonnez de magie*. La Haye: Adrian Vlac, MDCLIII.

Nendick, J., Scrancher, D., Usher, O. 2007. "Chlorine and Prout's hypothesis." H. Chang and C. Jackson, eds., *An Element of Controversy*. London: British Society for the History of Science, pp. 73–104.

Norton, B. 1978. "Karl Pearson and statistics: The social origins of scientific innovation." *Social Studies of Science* 8: 3–34.

Orléan, A. 2004. *Analyse économique des conventions*. Paris: Presses universitaires de France.

Pacioli, L. 1494. *Summa de arithmetica geometria, proportioni et proportionalita*, Venice: Paganino de Paganini.

Panofsky, E. 1967. *Architecture gothique et pensée scolastique*. Paris: Éditions de Minuit.

Paravicini Bagliani, A. 1975. "Witelo et la science optique à la cour pontificale de Viterbe, 1277." *Mélanges de l'École française de Rome* 87(2): 425–453.

Pasteur, L. 1951. *Correspondance de Pasteur*, réunie et annotée par Pasteur Vallery-Radot, tome II. *La seconde étape: fermentations, générations spontanées, maladies de vins, des vers à soie, de la bière (1857–1877)*. Paris: Flammarion [1940].

———. 1922. *Oeuvres de Pasteur*, réunies par Pasteur Vallery-Radot, tome II: *Fermentations et générations dites spontanées*. Paris: Masson.

———. 1862. "Mémoire sur les corpuscules organisés qui existent dans l'atmosphère. Examen de la doctrine des générations spontanées." *Annales de chimie et de physique* 3rd series, 64: 5–110.

Pecham, J. 1972. *Tractatus de perspectiva*, Edited with an Introduction and Notes by. D.C. Lindberg. New York: The Franciscan Institute.

Pedler, E. 1998. "Une sociologie historique des conditions d'existence de la musique." Weber, M. *Sociologie de la musique. Les fondements rationnels et sociaux de la musique*. Paris: Métailié, 155–195.

[Peisse, L.?] 1830a. "Lettre biographique sur l'école de Montpellier." *Gazette médicale de Paris* 18: 161–163.

———. 1830b. "Deuxième lettre biographique sur l'école de Montpellier." *Gazette médicale de Paris* 26: 229–231.

———. 1830c. "Troisième lettre biographique sur l'école de Montpellier." *Gazette médicale de Paris* 37: 335–339.

———. 1830d. "Quatrième lettre biographique sur l'école de Montpellier." *Gazette médicale de Paris* 45: 403–405.

Peisse, L. 1843. "Deuxième lettre à M. le Professeur Lordat." *Gazette médicale de Paris* 11(3): 37–42, (4): 53–59 (suite), (6): 85–93 (suite et fin).

———. 1841. "Philosophie médicale. Lettre à M. le Professeur Lordat." *Gazette médicale de Paris* 9: 113–119.

———. 1840. "Préface" to the *Fragments de philosophie* par M. William Hamilton. trad. L. Peisse. Paris: Madrange, pp. i–cxxxix.

Pennetier, G. 1907. *Un Débat scientifique. Pouchet et Pasteur*. (Actes du Muséum d'Histoire naturelle de Rouen, tome XI). Rouen: Imprimerie J. Lecerf.

Pestre, D. 1995. "Pour une histoire sociale et culturelle des sciences: nouvelles définitions, nouveaux objets, nouvelles pratiques." *Annales HSS* 50: 487–522.

Phillips, D.L. 1977. *Wittgenstein and Scientific Knowledge: A Sociological Perspective*. London: Macmillan.

Piaget, J. 1965. *Études sociologiques*. Genève: Librairie Droz.

Poincaré, H. 1970. *La Valeur de la science*. Paris: Flammarion [1905].

———. 1968. *La Science et l'hypothèse*. Paris: Flammarion [1902].

Polanyi, M. 1958. *Personnal Knowledge*. London: Routledge and Kegan Paul.

Popper, K.R. 1995. *La Théorie quantique et le schisme en physique*. Paris: Hermann.

———. 1975. *Objective Knowledge. An Evolutionary Approach*. Oxford: Oxford University Press. French translation: *La Connaissance objective*. Paris: Aubier, 1991.

———. 1963. *Conjectures and Refutations. The Growth of Scientific Knowledge*. London: Routledge. French translation: *Conjectures et réfutations. La croissance du savoir scientifique*. Paris: Payot, 1985.

———. 1945. *The Open Society and Its Enemies*, 2 vols. London: Routledge. French translation: *La Société ouverte et ses ennemis*, 2 vols. Paris: Le Seuil, 1979.

——. 1935. *Logik der Forschung. Zur Erkenntnistheorie der modernen Natur-wissenschaft.* Wien: Julius Springer. English translation: *The Logic of Scientific Discovery.* London: Routledge, 1992. French translation: *La Logique de la découverte scientifique.* Paris: Payot, 1973.

Pouchet, F.-A. 1873. *Catalogue des livres de la bibliothèque de feu M. Félix-Archimède Pouchet.* Rouen: Librairie A. Le Brument.

——. 1864. *Nouvelles Expériences de génération spontanée et la résistance vitale.* Paris.

——. 1859. *Hétérogénie, ou traité de la génération spontanée basé sur de nouvelles espériences.* Paris: J.-B. Baillière et fils.

——. 1853. *Histoire des sciences naturelles au Moyen Âge, ou Albert le Grand et son époque considérés comme point de départ de l'école expérimentale.* Paris: J.-B. Baillière.

Poudra, N.G. 1864. *Oeuvres de Desargues.* Paris: Leiber, 2 vols.

Quine, W.V.O. 1990. *The Pursuit of Truth.* Cambridge: The MIT Press. French translation: *La Poursuite de la vérité.* Paris: Éditions du Seuil, 1993.

——. 1970. *Philosophy of Logic.* Englewood Cliffs: Prentice-Hall. French translation: *Philosophie de la logique.* Paris: Aubier, 1975.

——. 1960. *Word and Object.* Cambridge: The MIT Press. French translation: *Le Mot et la Chose.* Paris: Flammarion, 1977.

——. 1953. *From a Logical Point of View.* Cambridge: Harvard University Press.

——. 1951. "The two dogmas of empiricism." *The Philosophical Review* 60: 20–43. French translation: "Les deux dogmes de l'empirisme, Domaine et langage de la science." P. Jacob, ed, *De Vienne à Cambridge.* Paris: Gallimard, 1980, 93–121, 219–240.

Quinn, A. 1984. *Logical Incrementalism.* New Haven: Yale University Press.

Radnitzky, G. 1990. "La révolution kuhnienne est-elle une fausse révolution?" *Archives de philosophie* 53(2): 199–212.

——. 1987. "La perspective économique sur le progrès scientifique: application en philosophie de la science du modèle coût-bénéfice." *Archives de philosophie* 50: 177–198.

Rapoport, A. 1960. *Fights, Games, and Debates.* Ann Arbor: University of Michigan Press. French translation, *Combats, Débats et Jeux.* Paris: Dunod, 1967.

Rashed R. 1999. *Les Mathématiques infinitésimales du IXᵉ au XIᵉ siècle*, III: *Ibn al-Haytham. Théorie des coniques, constructions géométriques et géométrie pratique*, London: al-Furqān.

——. ed. 1997. *Histoire des sciences arabes.* Paris: Éditions du Seuil, 3 vols.

——. 1974. *Condorcet. Mathématique et Société.* Choix de textes et commentaires. Paris: Hermann.

Raynaud, D. *See the selection of author's work at the end of the bibliography.*

Rescher, N. 2006. *Error.* Pittsburgh: University of Pittsburgh Press.

——. 1988. *Rationality.* Oxford: Oxford University Press.

———. 1978. *Scientific Progress: A Philosophical Essay on the Economics of Research in Natural Science.* Pittsburgh: University of Pittsburgh Press.

———. 1964. *The Development of Arabic Logic.* Pittsburgh: University of Pittsburgh Press.

Restivo, S. 2005. *Science, Technology and Society,* Oxford, Oxford University Press.

Rich, R.F. 1987. Politics, policy-making, and reaching closure. Engelhardt, H.T. and Caplan, A.L., eds., *Scientific Controversies.* New York: Cambridge University Press, 151–167.

Roger, J. 1890. *Les Médecins normands du XIIe au XIXe siècle,* tome I: *Seine-Inférieure.* Paris: G. Steinheil.

Roll-Hansen, N. 1980. "The controversy between biometricians and Medelians: A test case for the sociology of scientific knowledge." *Social Science Information* 19: 501–517.

———. 1979. "Experimental method and spontaneous generation: The controversy between Pasteur and Pouchet." *Journal of the History of Medicine and Allied Sciences* 34: 273–292.

Rozenfeld, B.A. and Yushkevich, A.P. 1961. "Dokazatel'stva pjatogo postulata Evklida u Sabita ibn Korry i Shams ad-Dina as-Samarkandi." *Istoriko-Matematicheskie Issledovania* 14: 587–592.

Rudwick, M.J.S. 1985. *The Great Devonian Controversy. The Shaping of Scientific Knowledge among Gentlemanly Specialists.* Chicago: The University of Chicago Press.

Sales-Girons, J. 1852. "Du vitalisme de l'école Montpellier. Prétentions de M. Lordat à l'orthodoxie religieuse." *Revue médicale française et étrangère* 15: 564–570.

Saurel, L. 1852. "Encore la Revue médicale!" *Revue thérapeutique du Midi* 3: 415–417.

Schantz, R. and Seidel, M., eds. 2011. *The Problem of Relativism in the Sociology of (Scientific) Knowledge.* Frankfurt am Main: Ontos.

Schatzman, E. 1989. *La Science menacée.* Paris: Odile Jacob.

Scheler, M. 1984. *Problèmes de sociologie de la connaissance.* Paris: Presses universitaires de France.

Schmitt, C.B. 1967. "Experimental evidence for and against a void: The Sixteenth-Century arguments." *Isis* 58: 352–366.

Scholz, H. 1961. *Concise History of Logic.* New York: Philosophical Library.

Schweber, L. 1997. "Controverses et styles de raisonnement." *Enquête* 5: 83–108.

Scott, P., Richards, E. and Martin, B. 1990. "Captives of controversy: The myth of the neutral social researcher in contemporary scientific controversies." *Science, Technology, & Human Values* 15(4): 474–494.

Serres, M., ed. 1989. *Éléments d'histoire des sciences.* Paris: Bordas.

Seymour, M. 2000. "Philosophie de la logique." P. Engel, ed. *Précis de philosophie analytique.* Paris: Presses universitaires de France, 119–141.

Shapin, S. 1979. "The Politics of observation: cerebral anatomy and social interests in the Edinburgh phrenology disputes." Wallis, R., ed., *On the Margins of Science*. Keele: University of Keele, 139–178. French translation: "La politique des cerveaux: la querelle phrénologique au XIXe siècle à Édimbourg." Latour, B. and Callon, M., eds., *La Science telle qu'elle se fait*. Paris: La Découverte, 1991, 146–199.

——. and Schaffer, S. 1989. *Leviathan and the Air-Pump: Hobbes, Boyle, and the Experimental Life*. Princeton: Princeton University Press. French translation: *Léviathan et la pompe à air, Hobbes et Boyle entre science et politique*. Paris: La Découverte, 1993.

Sharov, A.S. and Novikov, I.D. 1993. *Edwin Hubble: The Discoverer of the Big Bang*. Cambridge: Cambridge University Press. French translation: *Edwin Hubble. L'inventeur du Big Bang*, Paris: Flammarion, 1995.

Shinn, T. 1994. "Les dessous de la sociologie des sciences." R. Boudon and M. Clavelin, eds., *Le Relativisme est-il résistible?* Paris: Presses universitaires de France, 77–81.

——. 1980. "Division du savoir et spécificité organisationnelle." *Revue française de sociologie* 21: 3–35.

Shwayder, D.S. 1969. "Wittgenstein on mathematics." Winch, P., ed., *Studies in the Philosophy of Wittgenstein*. London: Routledge and Kegan Paul.

Siegel, H. 1987. *Relativism Refuted: A Critique of Contemporary Epistemological Relativism*. Dordrecht: D. Reidel Publishing Co.

Simmel, G. 1981. *Sociologie et Épistémologie*. Paris: Presses universitaires de France.

——. 1964. *Conflict and the Web of Group Affiliations*. New York: The Free Press [1908].

——. 1908. *Soziologie. Untersuchungen über die Formen der Vergesellschaftung*. Leipzig: Verlag von Duncker & Humblot. French translation: *Sociologie. Étude sur les formes de la socialisation*. Paris: Presses universitaires de France, 1999.

Simon, H.A. 1983. *Models of Bounded Rationality*, 2 vols. Cambridge: The MIT Press.

Smith, A.M. 2001. *Alhacen's Theory of Visual Perception*. A Critical Edition, with English Translation and Commentary, of the First Three Books of Alhacen's *De Aspectibus*, the Medieval Latin Version of Ibn al-Haytham's *Kitāb al-manāzir*, Philadelphia, American Philosophical Society.

Solère, J.L. 1995. "Antoine Arnauld ou la controverse dans les règles." A. Le Bouellec, ed. *La controverse religieuse et ses formes*. Paris: Cerf, 319–372.

Spade, P.V. 1977. "Roger Swyneshed's *Obligationes*: edition and comments." *Archives d'histoire doctrinale et littéraire du moyen âge* 44: 243–85.

——. and Stump, E. 1983. "Walter Burley and the *Obligationes* of William of Sherwood." *History and Philosophy of Logic* 4: 9–26.

Staszak, A. 1998. "Sociologie de la réception de Nietzsche." *L'Année sociologique* 48(2): 365–384.

———. 1994. *Les Usages de Nietzsche dans les sciences sociales en France. Étude sur la diffusion du nietzschéisme de 1889 à 1993.* Doctoral thesis. Paris: Université Paris 4.

Stengers, I. 1993. *L'invention des sciences modernes.* Paris: Éditions La Découverte.

———. and Schlanger, J. 1988. *Les Concepts scientifiques.* Paris: Éditions La Découverte.

Stoffel, J.F. (2002) *Le Phénoménalisme problématique de Pierre Duhem.* Bruxelles: Académie royale de Belgique.

Stroumsa, S. 1999. "Ibn al-Rāwandī's sū' adab al-mujādala: The role of bad manners in medieval disputations." H. Lazarus-Yafeh, M. R. Cohen, S. Somekh, S.H. Griffith, eds., *The Majlis: Encounters in Medieval Islam.* Wiesbaden: Harrassowitz, 66–83.

Stuewer, R.H. 1985. "Artificial disintegration in the Cambridge-Vienna controversy." Achinstein, P. and Hannaway, O., eds., *Observation, Experiment, and Hypothesis in Modern Physical Science.* Cambridge: Harvard University Press, 239–307.

Stump, E. 1985. "The logic of disputation in Walter Burley's treatise on obligations." *Synthese* 63: 355–374.

Suwaysī, M., ed. 1984. *Ashkāl al-ta'sīs li-'l-Samarqandī.* Tunis: al-Dār al-Tunīsiyya, 23–26.

Svensmark, H. 1998. "Influence of cosmic rays on earth's climate." *Physical Review Letters* 81(22): 5027–5030.

———. and Friis-Christensen, E. 1997. "Variation of cosmic ray flux and global cloud coverage: a missing link in solar-climate relationships." *Journal of Atmospherical and Solar-Terrestrial Physics* 59(11): 1225–1232.

Tachau, K.H. 1988. *Vision and Certitude in the Age of Ockam: Optics, Epistemology and the Foundations of Semantics 1250–1345.* Leiden: E.J. Brill.

Tahiri, H. 2008. "The birth of scientific controversies. The dynamics of the Arabic tradition and its impact on the development of science: Ibn al-Haytham's challenge of Ptolemy's *Almagest.*" S. Rahman, T. Street, and H. Tahiri, eds., *The Unity of Science in the Arabic Tradition.* Berlin/Heidelberg/New York: Springer, 183–224.

Tarski, A. 1933. "Projecie prawdy w jezykach nauk dedukcyjnych" (1933). German transl.: "Der Wahrheitsbegriff in den formalisierten Sprachen" (1936). English transl.: "The Concept of Truth in Formalized Languages." *Logic, Semantics, Metamathematics,* ed. J. Woodger. Oxford: Oxford University Press, 1956, 152–278.

al-Tawḥīdī 1970. *Kitāb al-imtā' wa-al-mu'ānasa,* edited by A. Amīn and A. al-Zayn. Beyrouth: Manshūrāt Dār maktabat al-Ḥayāt.

Theon of Alexandria 1895. *Euclidis Optica, Opticorum Recensio Theonis, Catoptrica, cum scholiis antiquis,* edidit I.L. Heiberg. Leipzig: B.G. Teubner.

Thorndike, L. 1923. *History of Magic and Experimental Science During the First Thirteenth Centuries of our Era,* vol. 2. New York: Macmillan.

Vajda, G. 1963. "Études sur Qirqisānī, V: Les règles de la controverse dialectique." *Revue des Études Juives* 122: 7–74.

Vandenbosch R. and Vandenbosch, S.E., eds. 2008. *Nuclear Waste Stalemate: Political and Scientific Controversies*. Salt Lake City: University of Utah Press.

Venetus, P. 1988. *Pauli Veneti Logica Magna*, Part II Fasc. 8: *Tractatus de obligationibus*, edited and translated by E.J. Ashworth. Oxford: Oxford University Press.

Venturini, T. 2010. "Diving in magma: How to explore controversies with actor-network theory." *Public Understanding of Science* 19(3): 258–273.

Vidal, Y. 1958. *La Bibliothèque et les Archives de la Faculté de médecine de Montpellier*. Montpellier: Causse, Graille et Castelnau.

Vinck, D. 1995. *Sociologie des sciences*. Paris: Armand Colin.

Vuillemin, J. 1986. "On Duhem's and Quine's thesis." Schilpp, P.A., ed., *The Philosophy of W.V.O. Quine*. La Salle: Open Court.

Walliser, B. and Zwirn, D. 1997. "Les règles de révision des croyances." Dupuy, J.P. and Livet, P., eds., *Les Limites de la rationalité*. Paris: La Découverte, vol. 1, 190–208.

Weber, M. 1921. *Wirtschaft und Gesellschaft. Grundriss der verstehenden Soziologie*. Tübingen: Mohr, 1976. English translation: *The Theory of Social and Economic Organization*. London: W. Hodge, 1947. French translation: *Économie et Société*. Paris: Plon, 1995.

———. 1904–1917. *Methodologische Schriften*. Frankfurt a.M.: S. Fischer, 1968. English translation: *The Methodology of the Social Sciences*. Glencoe: Free Press, 1949. French translation: *Essais sur la théorie de la science*. Paris: Plon, 1964.

Wittgenstein, L. 1958. *Preliminary Studies for the 'Philosophical Investigations'*. Oxford: B. Blackwell. French translation: *Le Cahier bleu et le Cahier brun*. Paris: Gallimard, 1965.

———. 1956. *Remarks on the Foundations of Mathematics*, ed. by G.H. von Wright, R. Rhees, G.E.M. Anscombe. Oxford: B. Blackwell. French transl.: *Remarques sur le fondement des mathématiques*. Paris: Gallimard, 1983.

———. 1953. *Philosophische Untersuchungen/Philosophical Investigations*, edited by G.E.M. Anscombe. New York: Macmillan. French transl.: *Investigations philosophiques*. Paris: Gallimard, 1961.

———. 1922. *Tractatus logico-philosophicus*. London: K. Paul. French translation by P. Klossowski. Paris: Gallimard, 1961. French translation by G.G. Granger. Paris: Gallimard, 1993.

Woolgar, S. 1988. *Science: The Very Idea*. London: Tavistock.

———., ed. 1988. *Knowledge and Reflexivity: New Frontiers in the Sociology of Knowledge*. London: Sage.

———. 1981. "Interests and explanations in the social study of science." *Social Studies of Science* 11: 365–397.

Xu, F., and Cheng, Z. 2009. "A tentative research on academical sovereignty in scientific controversies." *Studies in Dialectics of Nature* 25(5): 75–81.

Zemplén, G.A. 2008. "Scientific controversies and the pragma-dialectical model: Analysing a case study from the 1670s, the published part of the Newton-Lucas correspondence." F.H. van Eemeren, and B. Garssen, eds., *Controversy and Confrontation: Relating controversy analysis with argumentation theory*. Amsterdam: John Benjamin, 249–273.

A Selection of Author's Publications

Raynaud, D. "Un fragment du *De speculis comburentibus* de Regiomontanus copié par Toscanelli et inséré dans les carnets de Leonardo (*Codex Atlanticus*, 611rb/915ra)." *Annals of Science*, online 9 Sep 2014.

——. "A Tentative Astronomical Dating of Ibn al-Haytham's Solar Eclipse Record." *Nuncius. Journal of the Material and Visual History of Science* 29, 2014: 324–358.

——. "L'application des sections coniques au tracé de l'arc rampant par Nicolas-François Blondel," *Actes du Deuxième Congrès d'Histoire de la Construction* (Lyon, January 29–31, 2014). Paris: Picard (in press).

——. *Optics and the Rise of Perspective. A Study in Network Knowledge Diffusion*. Oxford: Bardwell Press, 2014.

——. "Building the Stemma Codicum from Geometrical Diagrams, A Treatise on Optics by Ibn al-Haytham as a Test Case." *Archive for History of Exact Sciences* 68–2 (2014): 207–239.

——. "Optics and Perspective prior to Alberti." B. Paolozzi Strozzi and M. Bormand, eds., *The Spring of Renaissance. Sculpture and Arts in Florence*. Florence: Madragora, 2013, 165–171.

——. "Les déterminations de la vitesse de la lumière (1676–1983). Étude de sociologie internaliste des sciences." *L'Année sociologique* 63–2 (2013): 359–398. English translation: "Determining the Speed of Light (1676–1983): An Internalist Study in the Sociology of Science," *L'Année sociologique/Cairn International* http://www.cairn-int.info/disc-sociology.htm.

——. "Leonardo, Optics, and Ophthalmology." F. Fiorani and A. Nova, eds., *Leonardo da Vinci and Optics*. Venice: Marsilio Editore, 2013, 255–276.

——. "Abū al-Wafā' Latinus? A Study of Method." *Historia Mathematica* 39 (2012): 34–83.

——. "Les débats sur les fondements de la perspective linéaire de Piero della Francesca à Egnatio Danti. Un cas de mathématisation à rebours." *Early Science and Medicine* 15 (2010): 474–504.

———. "La perspective aérienne de Léonard de Vinci et ses origines dans l'optique d'Ibn al-Haytham (*De aspectibus*, III, 7)." *Arabic Sciences and Philosophy* 19 (2009): 225–246.

———. "Le tracé continu des sections coniques à la Renaissance, applications optico-perspectives, héritage de la tradition mathématique arabe." *Arabic Sciences and Philosophy* 17 (2007): 299–345.

———. *La Sociologie et sa vocation scientifique*. Paris: Hermann, 2006.

———. "Duhem, Quine, Wittgenstein and the sociology of scientific knowledge." *Epistemologia* 26 (2003): 133–160.

———. "Ibn al-Haytham sur la vision binoculaire: un précurseur de l'optique physiologique." *Arabic Science and Philosophy* 13–1 (2003): 79–100.

———. "Effets de réseau dans la science pré-institutionnelle: le cas de l'optique médiévale." *European Journal of Sociology* 42–3 (2001): 483–505.

———. "La correspondance de F.-A. Pouchet avec les membres de l'Académie des Sciences: une réévaluation du débat sur la génération spontanée." *European Journal of Sociology* 40–2 (1999): 257–276.

———. "Les normes de la rationalité dans une controverse scientifique: l'exemple de l'optique médiévale." *L'Année sociologique* 48–2 (1998): 447–466.

———. "La controverse entre organicisme et vitalisme. Étude de sociologie des sciences." *Revue française de Sociologie* 39–4 (1998): 721–750.

———. *L'Hypothèse d'Oxford. Essai sur les origines de la perspective*. Paris: Presses universitaires de France, 1998.

Index rerum

Note: Scientific disciplines and noncontroversial theories and concepts are listed as main entries of this index. Disputed theories and discoveries are to be found under the Controversy entry

ethical or moral –, 3, 209
ideological or political –, 122, 126
philosophical –, 85
religious –, 145
Verification (and verificationism), 13, 19, 170, 172, 192, 194–5, 205, 208, 214–7, 219, 222

Wissensoziologie, 1–2

Zero, 4, 184

Index nominum